T0112412

THE SCIENCE OF
DOCTOR WHO

THE SCIENCE OF
DOCTOR WHO

The Scientific Facts Behind the Time Warps and Space Travels of the Doctor

MARK BRAKE

AUTHOR OF *THE SCIENCE OF STAR WARS, THE SCIENCE OF SUPERHEROES,* AND *THE SCIENCE OF SCIENCE FICTION*

Skyhorse Publishing

Copyright © 2021 by Mark Brake

All rights reserved. No part of this book may be reproduced in any manner without the express written consent of the publisher, except in the case of brief excerpts in critical reviews or articles. All inquiries should be addressed to Skyhorse Publishing, 307 West 36th Street, 11th Floor, New York, NY 10018.

Skyhorse Publishing books may be purchased in bulk at special discounts for sales promotion, corporate gifts, fund-raising, or educational purposes. Special editions can also be created to specifications. For details, contact the Special Sales Department, Skyhorse Publishing, 307 West 36th Street, 11th Floor, New York, NY 10018 or info@skyhorsepublishing.com.

Skyhorse® and Skyhorse Publishing® are registered trademarks of Skyhorse Publishing, Inc.®, a Delaware corporation.

Visit our website at www.skyhorsepublishing.com.

10 9 8 7 6 5 4 3 2

Library of Congress Cataloging-in-Publication Data is available on file.

Cover design by Daniel Brount
Cover image by Lauren Chisholm

Print ISBN: 978-1-5107-5786-8
Ebook ISBN: 978-1-5107-5787-5

Printed in China

To Emperor Whovian and Leader of the Robots of Death, Peter Grehan, the Doctor who never was, and whose mind is forever full of Daleks and Davros, Cybermen and Silurians, Master and Movellans

Contents

Introduction

"After all, in a world where very little is a surprise, and everything is viewed with cynicism, *Doctor Who* is a genuine rarity. It represents one of the very few areas where adults become as unashamedly enthusiastic as children. It's where children first experience the thrills and fears of adults, and where we never know the exact ending in advance. With its ballsy women, bisexual captains, working-class loquaciousness, scientific passion and unremittingly pacifist dictum, it offers a release from the dispiritingly limited vision of most storytelling. It is, despite being about a 900-year-old man with two hearts and a space-time taxi made of wood, still one of our very best projections of how to be human."

—Caitlin Moran, on the set of *Doctor Who*, *The Times* (2011)

I arrived on this planet with the newborn Space Age: these days of space exploration and our modern culture catalyzed by the discoveries of space. The Age began with Sputnik 1 in October 1957, and has been with us ever since. Little wonder that the Age should spawn *Doctor Who*—a television program about an alien called the Doctor who explores the Universe in a time-traveling spaceship. *Doctor Who* first appeared on BBC television at 5:16:20 GMT on Saturday, November 23, 1963. The program airing was eighty seconds later than scheduled, due to the assassination of John F. Kennedy the day before.

I've been a fan of *Doctor Who* ever since, from "The Daleks" to "The Day of the Doctor" and beyond. As luck would have it, I live just a half-hour's drive from Cardiff and BBC Wales, home of the *Doctor Who* reboot since 2005. As I also designed and validated planet Earth's first science and science fiction university degree program, I guess this makes me a kind of "scholar-fan." The 2005 regeneration of the Doctor presented a perfect opportunity to study the science of *Doctor Who*—not just the Doctor's adventures fighting Cybermen, Daleks, and the Master, but also his flirtations with Newton, Einstein, and Darwin.

I'm hoping that readers who adore *Doctor Who* might also enjoy thinking a little more scientifically about their favorite television show. In terms of ideas and metaphors, *Doctor Who* has infused our language and made it richer. It has certainly colonized British consciousness more markedly than the many alien races that have appeared on the show. And *Doctor Who* has gone global. Millions of fans across six continents enjoyed *The Day of the Doctor* in a worldwide simulcast and cinema extravaganza. Fans in over seventy-five countries—from Colombia to Canada, Botswana to Brazil, and Myanmar to Mexico—watched the fiftieth anniversary show on November 23, 2013, at the same time as the BBC One British broadcast.

In this book, my focus is on televised *Doctor Who,* from 1963 to date. There will only be passing references of the Doctor in other formats, such as tie-in novels, graphic art, or fan fiction. *Doctor Who* is science fiction, of course. But the tales of the Doctor are a peculiarly *British* kind of sci-fi. In 1999, British journalist Jeremy Paxman interviewed that other man who fell to Earth in David Bowie. During the course of the conversation, Paxman asked Bowie about the nature and identity of British rock music. Bowie's answer was fascinating, "We've always been good at music. We're not truly a rock nation. Everything we do in rock 'n' roll has a sense of irony attached to it. We know that we're not the Americans. We know it didn't spring from *our* souls. So, as the British always do, they try and do *something* with it, to make them feel smug. And that's what we're good at doing." It's the same in other parts of modern culture. When *RuPaul's Drag Race* was brought to the UK, contestants didn't adopt grandiose American names such as BeBe Zahara Benet, Pandora Boxx, or Sasha Velour, but far more ironically British

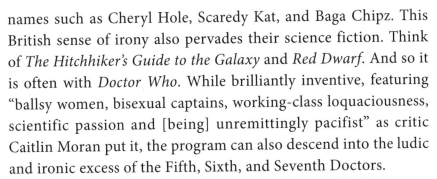

names such as Cheryl Hole, Scaredy Kat, and Baga Chipz. This British sense of irony also pervades their science fiction. Think of *The Hitchhiker's Guide to the Galaxy* and *Red Dwarf*. And so it is often with *Doctor Who*. While brilliantly inventive, featuring "ballsy women, bisexual captains, working-class loquaciousness, scientific passion and [being] unremittingly pacifist" as critic Caitlin Moran put it, the program can also descend into the ludic and ironic excess of the Fifth, Sixth, and Seventh Doctors.

If *Doctor Who* has been about one thing over all these years, it's been about the "weird." All science fiction is about the cultural shock of discovering our marginal position in an alien Universe. Sci-fi works by conveying the taste, the feel, and the human meaning of the discoveries of science. *Doctor Who* is an attempt to put the stamp of humanity back onto the Universe. To make human what is alien. Even in the form of the Doctor him (or her) self.

The weirdness of *Doctor Who* is also concerned with the relationship between the human and the nonhuman. *Doctor Who* seems to present an infinity of nightmares and visions. A bewildering array of conflicting themes: aliens and time machines, spaceships and cyborgs, utopias and dystopias, androids and alternate histories. But, on a more thoughtful level, we can identify four main themes: *space, time, machine,* and *monster.* Each of these themes is a way of exploring the relationship between the human and the nonhuman. Taking a closer look at these themes will enable a clearer understanding of the ways in which *Doctor Who* works, and what the program has to say about science.

Space

The space theme in *Doctor Who* sees the nonhuman as some aspect of the natural world, such as vast interstellar spaces in which the Doctor travels, or the alien, which can be seen as an animated version of nature. Here we look at the likes of space travel, baby Universes, and the science of exoplanets.

Time

This theme portrays a flux in the human condition brought about by processes revealed in time. Tales on time often focus on the dialectic of natural history, so they are of particular relevance to evolution and biology. Here we look at topics such as time machines, alternate histories, and regeneration.

Machine

Machine stories deal with the "man-machine" motif, including robots, computers, and artificial intelligences. Dystopian tales are part of the man-machine theme; it is the *social machine* in which the human confronts the nonhuman in such cases. This part has entries such as Daleks and Cybermen, superweapons, and *Doctor Who*'s very best invented machines.

Monster

Stories that focus on the nonhuman in the form of monster are usually situated within humanity itself. Especially if we read the Doctor as being essentially human as well as alien. In monster tales there is often an agency of change, such as nuclear war, which leads to the change of human into nonhuman, or Kaled into Dalek. It's within this theme that the remaking of humanoids through genetic design is encountered. Of course, monsters can be upbeat too, as the example of the Doctor as superhero testifies.

This way of thinking about *Doctor Who*, as the human versus the nonhuman, is pleasingly elegant and transparent. It helps us chart the Doctor's ongoing dialogue with science:

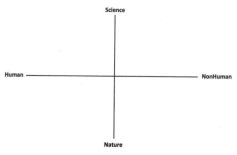

At times, with stories like "Sleep No More," science and the human in *Doctor Who* are pitched against nature and the nonhuman. In this case, the nonhuman comes in the form of a tectonic realignment that results in India and Japan becoming merged into Indo-Japan, and a new Indo-Japanese culture. In dystopias, such as "Gridlock" and "Turn Left," nature and human are united in opposition to science and nonhuman. As these dystopias suggest, science fiction may characterize science as nonhuman and unnatural. In "Gridlock," for example, the natural and organic human/alien combination of the Doctor and Martha counters the mechanical world of the perpetual gridlock within the Motorway, a highway system beneath the city state of New New York. According to sci-fi convention, utopias are imagined societies that are more fully human than the present.

More often, though, science features on both sides of the human-nonhuman conflict. In the many Dalek stories, for example, science is part of the nonhuman element symbolized by the invading Daleks. They are agents of the void. They also embody science with their vast, cool, and unsympathetic intellects. But the Daleks have to deal with the Doctor, a more human alien and a master exponent of science, who is always on the side of the invaded humans.

So welcome to *The Science of Doctor Who*, where the Doctor steps smoothly in and out of different realities, facing both earthly and unearthly threats with innovation and unpredictability, using science in the pay of nonviolent and intelligent resistance to succeed over brute force in ways that continue to be universally relevant.

—Mark Brake, 2020

Part I
Space

Introduction

"Throughout all his existence man has been striving to hear the music of the spheres, and has seemed to himself once and again to catch some phrase of it, or even a hint of the whole form of it. Yet he can never be sure that he has truly heard it, nor even that there is any such perfect music at all to be heard. Inevitably so, for if it exists, it is not for him in his littleness. But one thing is certain. Man himself, at the very least, is music, a brave theme that makes music also of its vast accompaniment, its matrix of storms and stars. Man himself in his degree is eternally a beauty in the eternal form of things. It is very good to have been man. And so we may go forward together with laughter in our hearts, and peace, thankful for the past, and for our own courage. For we shall make after all a fair conclusion to this brief music that is man."

—Olaf Stapledon, *Last and First Men* (1930)

One of history's greatest ever artists was Dutch master Hieronymus Bosch. Bosch lived between 1450 and 1516, but that didn't stop a time traveler like the Doctor running into him. In "The Hollows of Time," the Sixth Doctor examines an exhibit and is told it looks like a gargoyle created by Bosch. And the Eighth Doctor even claims to have met Bosch. Bosch's greatest work is the triptych oil painting on oak panel called *The Garden of Earthly Delights*. The painting is among the most intricate and enigmatic paintings of Western history, and filled with iconography and symbolism that have sparked debate for centuries since.

This masterpiece of Bosch's shows a mysterious and invented world, full of the kind of strange and daunting details one might find in *Doctor Who*. The imposing portrait features a man with a tree for a body, who gazes out from Hell, giant birds dropping fruit into the mouths of naked people, slithering creatures invading paradise, and a devil-bird that devours a man whole. What is the meaning of all this, the most famous of his paintings? Perhaps Heaven and Hell are not the destinations of your soul, but states of being that live inside you. No one knows for sure.

And yet the reverse of *The Garden of Earthly Delights* should intrigue we Whovians just as much. When the triptych's wings are shut, the design of the outer panels becomes visible. Rendered in green–gray grisaille, the panels lack color, only serving to further enhance the splendid color within. But the grisaille also speaks of a time before the creation of a Sun and Moon, which were formed, according to Christians, to "give light to the Earth." The outer panels depict the creation of the world. God is shown as a tiny figure, top left. Bosch shows God the creator with a Bible on his lap. The Earth below is encapsulated in a transparent sphere. It recalls the traditional depiction of the created world as a crystal sphere held by God. Earth hangs suspended in the cosmos, which is painted as an impenetrable darkness, whose only inhabitant is God himself.

Doctor Who could never have been written in Bosch's day. Bosch's medieval world had hard limits in time and space. The world was but six thousand years old, according to Bible scholars. And the idea of space was that inherited from Aristotle. Aristotle's cosmos was a two-tier, geocentric Universe. The Earth, mutable and corruptible, was placed at the center of a nested system of crystalline celestial spheres, from the sub-lunary to the sphere of the fixed stars. The sub-lunary sphere—from the Earth to the Moon—was alone in being subject to the horrors of change, death, and decay. Beyond the Moon, the supra-lunary or celestial sphere, all was immutable and perfect. Crucially, the Earth was not just a physical center. It was also the center of motion, and everything in the cosmos moved with respect to this single center. Aristotle declared that if there were more than one world, more than just a single center, the elements of earth and fire would have more than one natural place toward which to move—in his view a rational and natural contradiction. Aristotle concluded that the Earth was unique. There was no room for the alien.

Barely a hundred years after Bosch painted *The Garden of Earthly Delights*, Aristotle's crystalline Universe was shattered by the telescope. And so the Scientific Revolution began, and along with it, science fiction. The demise of the days of Aristotle and Bosch marks the paradigm shift of the old Universe into the new. Their cozy geocentric cosmos was all about us humans. But the new Universe, the one that the Doctor inhabits, is decentralized, inhuman, infinite, and alien.

When you think about it this way, you can see *Doctor Who* as a way of coming to terms with an entirely different geographical space. And the Doctor's stories as a response to the cultural shock created by the discovery of our new and marginal position in a Universe fundamentally inhospitable to humans. *Doctor Who* is an attempt to make human sense of the new Universe.

So the science fiction of the Doctor's Whoniverse tells tales about what it's like to be human in the modern world. Before the days of science, stories like those in *Doctor Who* would have been impossible to write and imagine. When you live in Bosch's world, where the Universe was smaller than the distance from the Earth to the Sun, there's not so much to explore. And nobody out there but God.

Lots of the Doctor's tales are about the urge to escape the confines of space, of planet Earth, which in some ways is our prison. In one tale, the Third Doctor was actually kept prisoner on the Earth by the Time Lords. So, *Doctor Who* tells weird and wonderful tales to match the modern Universe we live in. As the Eleventh Doctor said, "All of time and space; everywhere and anywhere; every star that ever was. Where do you want to start?!"

What kind of space stories are found in *Doctor Who*? In short, stories about humans versus something nonhuman. In other words, humans meet some feature of space that allows them to discover and explore things. It could be weird planets, weird stars, or weird aliens. One *Doctor Who* writer once said the Doctor's stories were like galaxy and chips. Chips are very human (especially if you're British) and galaxies are very cosmic and nonhuman. And when someone asked Peter Capaldi what it was he loved about *Doctor Who* stories, he said, "It is this relationship between the domestic and the epic. The sense that there's a bridge, that a hand can be extended, and you can step from the Earth, from the supermarket car park, into the Andromeda nebulae or whatever."

So that's the theme of space. It's the void into which the Doctor hurtles in his TARDIS. And it's your turn to hurtle into this part off space, where the Whoniverse presents us with baby Universes, weird and wonderful exoplanets, and the strangest alien creatures you can imagine.

What Does *Doctor Who* Tell Us about Space Travel?

In the Fourth Doctor story "The Ark in Space" (1975), the TARDIS materializes on an aged space station, and the Doctor realizes the ark is a generation starship.

> There is no way back into the past; the choice, as Wells once said, is the Universe or nothing. Though men and civilizations may yearn for rest, for the dream of the lotus-eaters, that is a desire that merges imperceptibly into death. The challenge of the great spaces between the worlds is a stupendous one; but if we fail to meet it, the story of our race will be drawing to its close.
>
> —Arthur C. Clarke, *Interplanetary Flight* (1950)

The Ark in Space

Roughly forty-five light-years from planet Earth sits a star in the famous constellation Ursa Major, the Great Bear, whose associated mythology likely dates back into prehistory. The star in question is 47 Ursae Majoris, formally named Chalawan. What makes 47 Ursae Majoris so special? In 1996, it became one of the first stars to have exoplanets discovered in orbit about it, and it is now thought to have at least three known planets. The light that reaches Earth from 47 Ursae Majoris is forty-five years old, given the distance between our two systems.

And when that light was leaving 47 Ursae Majoris, a *Doctor Who* story called "The Ark in Space" was televised for the first time. In the 1975 tale, the TARDIS materializes on an old space station. When the

Fourth Doctor explores the old station, he realizes it's a kind of ark. The idea of such arks, or generation starships as they're sometimes known, is a good example of the way in which science and science fiction influence one another. That's because many of the engineers and scientists who wrote about generation starships were also sci-fi writers. One of the earliest accounts of a "space ark" is in the 1929 essay *The World, the Flesh, and the Devil*. It was written by Irish scientist J. D. Bernal. Bernal's essay was about the idea of human evolution and the human future in space. He described ways of living in space that included what we now call generation starships.

Pioneering rocket engineers also wrote about long-duration interstellar journeys. American rocket scientist Robert Goddard was among the first in his 1918 paper, *The Last Migration*. Here Goddard describes the death of our Sun and the journey of an "interstellar ark," on which its crew would travel for centuries in suspended animation until waking at their ultimate destination—another star system.

Russian rocket pioneers were at it too. Konstantin Tsiolkovsky, the father of astronautic theory, wrote an account of the need for multiple generations of passengers in his 1928 essay, *The Future of Earth and Mankind*. Tsiolkovsky described a cosmic "Noah's Ark"—a space colony kitted out with engines, which travels for thousands of years.

The Future Possibilities of Spaceflight

Human imagination is often light-years ahead of practice, of course. And, as the Doctor knows, we humans don't as yet have too much experience of spaceflight. Space arks are one thing, but so far humans have only traveled to the Moon. Our Galaxy, the Milky Way, is about one hundred thousand light-years wide. And destination Moon, to which we've sent a number of manned missions, is a mere light-second away. You can see the distance problem with space travel quite easily.

Consider speed. At the moment, space travel is slow and expensive. In the Fifth Doctor story "Enlightenment," the Doctor is aboard a spaceship that uses solar sails. In our real Universe, the fastest of these sail-ships work by aiming lasers (essentially beams of light) at specially made sails. When the laser hits the sails, it causes the ship to move more quickly than if they relied on the Sun alone.

Now, imagine we could launch such a solar-sailed ship from Earth. And imagine also that the ship aims a high-powered laser beam at the sails to propel it. That ship would take around forty years to reach the nearest stars. Not exactly quick, is it? Imagine how long a *Doctor Who* story would be at those speeds! And what about the stars and galaxies beyond? Now you can see why, from the very get-go, rocket pioneers and sci-fi writers realized humans would need generation starships to last long journeys.

Traveling the Whoniverse

Luckily, Whoniverse travel doesn't depend on solar sails. And, generally speaking, space travel is thought to be one of the first steps a species takes in their development as a sophisticated civilization. Like us, the Time Lords realized that the vacuum of space is an extremely hostile place that can seriously mash up your genes. Earth scientists think that a mission to Mars is just about doable for humans, as the radiation sickness would be deadly if we tried to venture out any farther. The main problem is "galactic cosmic rays"—the nuclei of atoms traveling at ultra-high speeds. One solution to such rays could be the one used in the *Doctor Who* episodes "42" and "The Impossible Planet." By the forty-second century, it seems that many human spaceships and bases have developed force fields to protect them from the harmful effects of space.

Rockets and spaceships are one way, but what other ways of space travel does *Doctor Who* suggest? Well, in the early *Doctor*

Who story "The Seeds of Death," we learn how humans stopped making rockets for spaceflight because of a new technology known as T-Mat. T-Mat stands for Travel Mat Relay. Using T-Mat tech, humans could be instantly teleported between the Earth and the Moon. But when humans stopped investing in rockets, we also lost interest in traveling beyond the Moon, then soon even forgot how to use rockets. Except, that is, for one guy who worked in a museum and was privately building his own retro rocket.

In later stories, the T-Mat tech is barely mentioned. Next, all we hear about is the Transmat, or matter transmitter. Transmat is used to transport humans between planets or space stations. And yet in one story a Transmat beam is used by the Time Lords to teleport the Doctor all the way to Skaro from future Earth. Transmat had considerable potential! So, according to the Time Lords, instantaneous space travel was possible. And before you say this is all nonsense, do remember that science is never finished. When so-called experts tell you that certain aspects of Whoniverse travel are impossible, remember to take their advice with a grain of salt. Science is always refining what we know. And if something seems impossible now, that doesn't mean it'll be impossible in the future.

Revisiting the Space Ark

Having said that, American sci-fi writer Kim Stanley Robinson revisited the question of the generation starship in a 2016 *Scientific American* article called "What Will It Take for Humans to Colonize the Milky Way?" Robinson's novel *Aurora*, published in 2015, follows a gargantuan generation ship and its seven generations of humans as they make their way to the Tau Ceti system, twelve light-years away, to start a human colony.

And what did Robinson have to say on the topic of space arks? "The idea that humans will eventually travel to and inhabit other parts of our galaxy was well expressed by the early Russian rocket scientist Konstantin Tsiolkovsky, who wrote, 'Earth is humanity's cradle, but you're not meant to stay in your cradle forever.'" But Robinson's conclusion was cagey.

An interstellar voyage would present one set of extremely difficult problems, and the arrival in another system, a different set of problems. All the problems together create not an outright impossibility, but a project of extreme difficulty, with very poor chances of success. The unavoidable uncertainties suggest that an ethical pursuit of the project would require many preconditions before it was undertaken. Among them are these: first, a demonstrably sustainable human civilization on Earth itself, the achievement of which would teach us many of the things we would need to know to construct a viable mesocosm in an ark (a mesocosm is an outdoor experimental system that examines the natural environment under controlled conditions. In this way mesocosm studies provide a link between field surveys and highly controlled laboratory experiments.) Second, a great deal of practice in an ark orbiting our sun, where we could make repairs and study practices in an ongoing feedback loop, until we had in effect built a successful proof of concept; third, extensive robotic explorations of nearby planetary systems, to see if any are suitable candidates for inhabitation. Unless all these steps are taken, humans cannot successfully travel to and inhabit other star systems.

What Spaceships Are "Fit" to Sail the Whoniverse?

In the Eleventh Doctor story "The Time of Angels" (2010), the Doctor journeys to the planet Alfava Metraxis, where the spaceship Byzantium has crashed. Hidden inside is a Weeping Angel.

"First, inevitably, the idea, the fantasy, the fairy tale. Then, scientific calculation. Ultimately, fulfilment crowns the dream . . . For me, a [spaceship] is only a means, only a method of reaching the depths of space, and not an end in itself . . . There's no doubt that it's very important to have [spaceships] since they will help mankind to settle elsewhere in the Universe. But what I'm working for is this resettling . . . The whole idea is to move away from the Earth to settlements in space."

—Konstantin Tsiolkovsky (1903) in *The Investigation of Universal Space by Means of Reactive Devices,* NASA translation (1964)

TARDIS

Well, of course we're going to start with the TARDIS. The now obsolete Type 40 TT Capsule, of which there were only 305 registered on Gallifrey, has been serving the Doctor for centuries of phone box travel. The ship was meant to have six pilots, but the Doctor just about gets by on his own. (Well, most of the time.) According to the Twelfth Doctor episode "Flatline," if the TARDIS were ever to land with its true weight, it would fracture the surface of the Earth.

Ship's appearance:
The TARDIS may change its appearance the millisecond it lands, in order to better fit with its environment. But the "chameleon circuit," responsible for the change, got broken, so the blue police phone box look is here to stay. (The chameleon circuit breakdown was a way to cut down costs and not to have to build a new TARDIS exterior set for every new story. It's similar to the USS *Enterprise* in *Star Trek*, which was originally envisaged as splitting into a saucer section, which then landed wherever. Too expensive and time-consuming for every new planet, so they famously came up with the transporter idea.)

Real-life probability:
Well, the actual phone boxes exist, though the BBC, shifty devils, actually came to own the copyright of the London police box design, so the London police couldn't build any more without BBC permission!

Genesis Ark

The Genesis Ark was a prison ship, created by the Time Lords during The Last Great Time War, which held millions of Daleks. Kind of useful and scary at the same time, the Ark had a life support system and a navigation engine. But the Daleks stole it, brought it to Earth, and released millions of Daleks into the skies above London. That surely can't be good.

Ship's appearance:
The Ark, according to Dalek Sec, needed an area of thirty square miles.

Real-life probability:
Unlikely.

Racnoss Webstar

The Webstar was a starship made of silk and strengthened by dark energy. It's not everyone's first thought when considering travel across the cosmos, even if it is the Whoniverse. Each Webstar could be steered by a single Racnoss, with the pilot being able to teleport to and from the surface of a planet, merely adding to the sheer improbability of the whole affair.

Ship's appearance:
An eight-pointed webbed star, looking very much like something a spider created in the vacuum of space.

Real-life probability:
Unlikely.

Judoon Rocket

This spacecraft was used by the Judoon to capture alien criminals. When the rocket lands, a long exit ramp appears, and huge numbers of Judoon soldiers troop out onto the unsuspecting planet. Powered by pulse fusion engines, Judoon rockets can cross the solar system in forty-five minutes. That's pretty fast.

Ship's appearance:
Classic-looking rockets with four landing prongs and four boosters.

Real-life probability:
More likely. Nuclear pulse propulsion is a hypothetical method of spacecraft propulsion that uses nuclear explosions for thrust. As to crossing the solar system in forty-five minutes: assuming the solar system has a diameter of over seven billion miles, the Judoon rocket would have to be traveling at 3.37 million miles a second. Unlikely.

Slitheen Craft

The spaceship that famously crashed into London's Big Ben, the Slitheen craft was a faster-than-light ship, which used a slipstream engine to work its way through space (and crash into Big Ben).

Ship's appearance:
Looking much like a fat, grey Raxacoricofallapatorian frisbee.

Real-life probability:
You do the "math."

Dalek Void Ship

Okay, for sheer cool, this spacecraft is some concept. Designed by the Cult of Skaro, its mission was to explore the void that exists between the dimensions of space. Sure, no one knows a lot about such voids, but that's no doubt the point of the mission. Allegedly, the ship could separate from time and space, which it would need to get into those all-important voidy bits. Finally, as it has no detectable mass or heat, it can become invisible. Apparently.

Ship's appearance:
Invisible.

Real-life probability:
Nope. Seems someone just liked the sound of "Dalek Void Ship" and went with it.

SS *Madame de Pompadour*

The SS *Madame de Pompadour* is a fifty-first-century energy trawler, which collects dark matter in its rotating arms that also double up as the ship's method of propulsion. The *Pompadour* starship has a crew of around fifty people, and was also famous for its deadly keen crew. Numbering also about fifty, the crew of clockwork repair droids would be ready to fix the ship at any cost to the crew, or themselves.

Ship's appearance:
Much like a space station, only far creepier.

Real-life probability:
As likely as Trump winning the Nobel Prize for Literature.

Sycorax Ship

The Sycorax Armada was the fleet of asteroid ships used by the Sycorax race for space travel. Literally built from asteroids, these city ships were so huge that they created shockwaves when they entered the Earth's atmosphere, and eclipsed the Sun.

Ship's appearance:
Like gigantic flying rocks. And watch out, their shockwaves will shatter all the glass in your home.

Real-life probability:
This one's not so improbable. Scientists and engineers dream of moving asteroids in the near future when the day comes to mine them for their mineral wealth. Having said that, no one is yet talking about turbocharging those babies.

Sontaran Flagship

The Flagship forms part of the Sontaran attack fleet. Around its "waist" is a belt of dozens of smaller Sontaran scout ships, spherical crafts accommodating one soldier each. The Flagship is resistant to nuclear weapons and carries a planet-ravaging cannon, revealed when the ship slips into battle mode (Sontarans are far from subtle).

Ship's appearance:
Like a cosmic flying bug—a large sphere with eight humongous claws; four on the top half and four on the bottom.

Real-life probability:
Nope.

The Ark

The Ark is a gigantic Generation Starship created millions of years into the future. It's on a 700-year mission to relocate the entire population of the dying Earth; including humans, animals, and a one-eyed race of aliens called Monoids. While it has an onboard city and a huge jungle, most of the humans are actually held at "microcell size" within storage trays, with the exception of a crew of human guardians and Monoids.

Ship's appearance:
We only ever see a portion of the ship's exterior, but a map suggests that it's a circular ship with a huge metallic roof. Basically, then, a flying shed.

Real-life probability:
According to American sci-fi writer Kim Stanley Robinson, pretty tricky.

Could the TARDIS Really Be Bigger on the Inside?

In the Eleventh Doctor story "The Vampires of Venice" (2010), the Doctor is a little disappointed when Rory is not as stunned as people usually are on entering the TARDIS inner space.

The TARDIS, Again!

The TARDIS is arguably history's most famous time machine. Camouflaged as a police box, and hiding a myriad of interior marvels, she has traveled the cosmos over, and often saved the Doctor, as well as the day. If you're ever in any doubt as to the importance of the TARDIS, just remember how nervous you feel watching *Doctor Who* when the TARDIS is trapped. It's unnerving! You realize at once that the Doctor's escape route through space-time may be blocked, and the Doctor himself left stranded. (Incidentally, the trapped TARDIS was an important plot device, especially in the early years of the program. The easy escape route needed to be cut off so that the Doctor and his companions were forced to solve the problem in which they found themselves.)

The exterior of the TARDIS has long been familiar to audiences about the globe. The interior, however, has changed almost as much as the Doctor himself. Not just the regular updates to her console room, but those rather rare times when we slip past her interior doors and delve into the endless, cavernous corridors within. The TARDIS has been the scene for some of *Doctor Who*'s most famous moments. Perhaps the most notable of these is the wonderfully dark tale from Neil Gaiman, *The Doctor's Wife*. In this Eleventh Doctor episode, we finally get a glimpse of the consciousness of the ship. After years of the Doctor treating her as a living thing, she is channeled into the body of a woman and the shell of the machine is stolen.

Bigger on the Inside

And yet the most often repeated sub-scenes involving the TARDIS are those many times when someone new is introduced to her "bigger on the inside" technology. To be clear, the TARDIS isn't the only thing bigger on the inside. There's the carpetbag belonging to Mary Poppins. Mary is able to keep all manner of objects in the bag, some much bigger than the bag appears to be, including a hat stand and potted plant. (We shall ignore here the many rumors that Mary Poppins is herself a Time Lord.) You get similar scenes in many cartoons, naturally, but for many folk Mary's bag is their first exposure to the confounding concept of bigger on the inside (if they hadn't seen the TARDIS!).

Then there's the Weasleys' tent from the Harry Potter books and movies. It looks like a two-person pitch at first glance, but on entering you discover a fully furnished canvas palace, complete with kitchen, bathroom, and bedrooms.

And don't forget the Phantom Zone. This was a clear, flat shard of an extra-dimensional prison in *Superman*. Three dangerous, traitorous criminals are weirdly imprisoned in the shard, and set to drift in space-time by Superman's Kryptonian father, Jor-El.

But it's the TARDIS that usually tops most people's list of objects bigger on the inside. It has a control room, for heaven's sake. It houses myriad other rooms, including the Doctor's quarters, an art gallery, and a swimming pool, until it sprung a leak. (The TARDIS also appears to be adjustable, being reconfigured according to which Doctor is in residence, which also suggests a psychological link.)

Bending Space-time

So how might the Time Lords have pulled off this "bigger on the inside" technology? One possible way is using gravity. The human understanding of gravity has a long history, running from Aristotle, through Newton, to Einstein. And it's Einstein's theory of gravity that we hold to today. It's the idea that gravity is a bending of space and time. So, for example, the mass of our Sun creates a gravity field, which makes a bowl-shaped dent in the space-time around it. The planets in orbit around the Sun are then somewhat like round chocolate drops, rolling around the inside of the bowl.

Now, if the Time Lords made the TARDIS out of the right kind of stuff, they could then use this gravity-bending to build a "bubble" that's bigger on the inside than out. The stuff they would use would be a very exotic type of matter, but hey, these Time Lords had been going for billions of years, so they probably know their way around the Whoniverse.

Imagine a spider (let's make it a Gallifreyan spider) crawling along a flat two-dimensional wall. Unknown to the spider, there is a "mouth" hidden in this wall. The "mouth" is attached to a "lobe," which looks a lot like a balloon. So in fact the narrow "mouth" opens out into a much bigger area. And if you scale this figure up into 3D, you roughly have the same situation as the TARDIS. The main problem is that the material you'd need to use would be "exotic matter." It's weird stuff. If you pumped your car tires with this "exotic matter," your tires would get flatter!

We witnessed just how odd TARDIS space can be in the Twelfth Doctor episode "Flatline." In the tale, the outside of the TARDIS shrinks, but everything inside, including the Doctor, remains the same size. This leads to a compelling scene where the Doctor's hand pops out of the TARDIS to try and drag his ship off the track of a looming train. Now try figuring *that* one out!

What Has *Doctor Who* Done with Darwin's Natural Selection?

In the Ninth Doctor story "Aliens of London" (2005), the alien crime family the Slitheen fake a spaceship crash-landing in the River Thames. They lure experts of extraterrestrial life, including the "ultimate expert" in the Doctor, into a trap inside 10 Downing Street.

> "My eyes are constantly wide open to the extraordinary fact of existence. Not just human existence, but the existence of life and how this breathtakingly powerful process, which is natural selection, has managed to take the very simple facts of physics and chemistry and build them up to redwood trees and humans."
>
> —Richard Dawkins, *The Guardian* (2013)

Darwin and the Doctor

Charles Darwin invented the modern alien. That great discovery of the nineteenth century, in which Darwin played a major part, was not just a theory that life evolved. That had been argued before. No, the true innovation of Darwin's age was to discover the evolutionary mechanism by which new species came about. Little wonder Darwin named his 1859 book, *The Origin of Species*. "Evolution" was not a word Darwin liked to use. "One may say," Darwin wrote, "there is a force like a hundred thousand wedges trying to force every kind of adapted structure into the gaps in the economy of nature, or rather forming gaps by thrusting out weaker ones."

Here was a natural mechanism that was not only global, but also cosmic. Extraterrestrial as well as terrestrial. It was a mechanism that canvassed continually to snuff out most traits. It kept only those traits carried by individuals who had won the struggle to survive and breed. Here was "natural selection." The individual differences between members of a species, along with environmental factors, shape the chances that an individual will pass its traits on to posterity.

What have the writers of *Doctor Who* done with this evolutionary mechanism over the last five decades or so? Well, here's the best point about all those aliens in *Doctor Who*. Even though science has made tremendous gains in the understanding of space during the twentieth and twenty-first centuries, scientists still have relatively little to say about the psychology and physiology of the alien. That's mostly still the job of science fiction. Sci-fi has been conducting a kind of continuous thought-experiment on the matter of aliens for a few centuries. And *Doctor Who* has been doing so for over half a century.

So let's take a look at the last five decades of *Doctor Who* aliens. A word of warning: you won't find the Silurians. Even though the Third Doctor referred to the Silurians as aliens, he soon learned they had ruled on planet Earth millions of years before.

Daleks

Late in their evolution, of course, the Daleks would hide deep down in their robot shells. And yet, inside those shells are living creatures, evolved and mutated from aliens known as the Kaleds, one of the humanoid races that originally inhabited the planet Skaro. The "mad doctor" Davros engineered the creatures to believe every other race inferior and to become hell-bent on conquering the Whoniverse. Dalek-creator Terry Nation based his invention on the Nazis. And when you watch the Daleks armed with this knowledge, their behavior makes a lot more sense! As Peter Grehan points out in his excellent 2016 book *Connecting Who: Artificial Beings*, in the very first two stories, the Daleks even make the Nazi salute. In "Genesis of the Daleks," the uniform of the Kaleds is very SS-like, and Nyder's badge had to be removed because it was too similar to the Nazi Schutzstaffel (SS) badge. Davros, incidentally, is the leader of a scientific elite. They have gradually usurped the power of the legitimate Kaled government, with the support of a quasi-military organization that wear black uniforms bearing insignia reminiscent of lightning bolts, and who salute each other by raising a hand, palm outward, and clicking the heels together—ring any bells? And a final clue, should there be any doubt: Nyder, the bespectacled commander of this organization, wears an Iron Cross (although this was removed for subsequent episodes). In short, Nyder is a parallel to Heinrich Himmler, Reichsführer of the SS, and Davros, a brilliant madman who, having maneuvered himself into a position of power, is obsessed with the creation of a "master race," is Hitler, while the Daleks themselves are composite creatures of various Nazi characteristics.

Weeping Angels

The Weeping Angels are a fascinating example of sci-fi writers going mad with the evolutionary mechanism idea. In this case that mechanism is a defense tactic known as quantum-locking. The Weeping Angels are meant to be an ancient race that dates back to the beginning of the Whoniverse. How did they last that long? This is where quantum-locking comes in. They move silently and quickly to kill their prey. When observed by any living creature (including other Weeping Angels), they turn to stone, making them resistant to harm. When they're not being observed, they can come to life; quickly! That's why it's not a good idea to even blink anywhere near one of these guys. (Another quota of evolutionary quirk is the idea that the Angels dump their victims by zapping them across time, so they can feed off the time energy. Whatever *that* is.) Actually, thinking once again about the quantum-locking mechanism: Does this mean Weeping Angels have to breed blindfolded?

Cybermen

To be truly extraterrestrial you merely need to be from a planet other than Earth, even if that planet is Mars or Earth's fictional twin planet Mondas. Mondas was the original home of the Cybermen. Like us, they're human in form, but they've been stripped of all emotion, and enhanced by cybernetic body parts. And yet, they have a weakness. The element gold is the Cybermen's Kryptonite. It plays havoc with their breathing, or something. For example, one of the Cybermen had his tongue pierced and paid the price by choking to death. (Actually, I just made up that last fact, but it's true about gold being their Kryptonite.)

The Silence

Another impossibilist evolutionary mechanism idea comes in the form of the Silence. Sure, they're creepy. And sure they're so obviously modeled on Edvard Munch's 1893 painting *The Scream*, famous for its agonized face, which has become one of the most iconic images of art, seen as symbolizing the anxiety of modern man. But the creepiest thing about the Silence is the fact they've been living alongside humans for thousands of years. And yet no one has ever remembered seeing them. In a kind of reverse-engineer of the quantum-locking mechanism of the Weeping Angels, as soon as you turn your back or look away from the Silence, you forget about them. Evolutionarily speaking, we must assume this mechanism doesn't work on the Silence themselves, otherwise they'd have had a troubled history in social gatherings.

Vashta Nerada

Possibly a more believable alien race are the Vashta Nerada. They're microscopic critters that live in shady swarms, casting creepy shadows when they emerge into the light. They live on almost every planet that has meat, so that includes you and yours. (One assumes they get from planet to planet inside space-faring human meat, which is also a rather creepy idea.) The story goes that they first emerged as tiny spores in trees, larking about in forests, but soon got bored and decided instead to strip your body down to bare bone. And just in case you think the Vashta Nerada are a fictional exaggeration, consider the parasitic wasp. Also known as the jewel wasp, the parasitic wasp goes up to an insect and, by imparting a sting into the brain, turns it into a kind of zombie. The wasp then pulls the insect into the wasp's nest, and wasp eggs are planted in the sealed nest. When the baby wasps are born, they eat the insect alive from the inside, in a special order, to keep it alive for as long as possible, so that the insect meat doesn't fall off too quickly.

The Ood

The curious-looking Ood are another humanoid alien. They have no voice, but they can communicate instead by the power of thought. One can only wonder in amazement as to how this little ability evolved. The Ood also have two brains. One brain is in the head (normal). The second brain they hold in their hands; it's connected by a kind of cord to their faces. Let's just say: highly improbable and laughably impractical. What's the evolutionary advantage of the opposable thumb if you have to *carry* your brain wherever you may roam? (More trivial and juvenile questions include: Do they munch Ood food? Is their family unit known as an Ood brood? And, when they get grumpy, are they in an Ood mood?)

Slitheen

The Slitheen are the most famous family of farting aliens in the Whoniverse. These Raxacoricofallapatorians are rotund of habit, eight feet tall, and have powerful claws. (The powerful claws are great for hunting, fighting, and generally scaring weaker species, but not so good for constructing intricate high-tech devices, changing a tire, or making sandwiches.) They are able to hide in the skins of humans, and other victims, by using a gas collar, worn around their necks, which allows them to fart out any excess gas (it's usually a tight squeeze, slipping into a human skin). One assumes that, before they had the necessary tech to invent the gas collar, the Slitheen simply bound about on Raxacoricofallapatorius, without the need for farting tech. (Incidentally, the Slitheen's gas collar looks remarkably similar to the patch on the tummy of the Teletubbies, though everyone seems to be mysteriously quiet about that . . .)

Sontarans

These three-fingered, toad-faced humanoid aliens are a militaristic species dedicated to a warring life. Being clones, they were batch-created in millions, and on a planet with more gravity than Earth's. Thus the Sontarans are short but strong. (They're also not very funny and totally devoid of humor. Unlike many species they also wear helmets that are exactly the same shape as their heads, somewhat like a jelly mold.) Like the Cybermen, the Sontarans also have a curious Kryptonite. Their weak spot is this: they can easily be stunned by a blow to a vent at the back of their necks. This makes it very advisable that they should always stay face-on to enemies in battle.

Which *Doctor Who* Aliens Should Never Have Made the Science Cut?

In the Tenth Doctor trailer of 2005, the new season of the Doctor promised that, outside the doors of the TARDIS, we might meet anything: new worlds, terrifying monsters, impossible things. And if we follow the Doctor, nothing will ever be the same again.

The Science of Aliens touring exhibition that launched at the London Science Museum in October 2005 asked the question "are we alone in the Universe?" The exhibition used science fiction archetypes as well as looking at what scientists can tell us about the real possibilities for alien life. A variety of experts gave advice on the exhibition development including Simon Conway Morris, Jack Cohen, and myself.

Writers had imagined aliens long before the days of *Doctor Who*, or even Darwin. The culture of imagining the unimaginable life beyond this Earth has a long history, going back two thousand years. At the heart of this alien fiction has been the desire to portray the true alien. Writers tried to imagine beings that are blessed with reason, but are not human. After Darwin, and his mechanism of natural selection, many writers relied too heavily on the influence of H. G. Wells, whose invading Martians in 1898's *The War of the Worlds* were bent on cosmic domination and the conquest of the Earth in particular.

But H. G. Wells's very specific scenario was copied *ad nauseam*. Science fiction stories trotted out mechanical simulacra of the Wellsian prototype. And audiences were served up a farcical parade of alien monstrosities, stretching believability to breaking point. In fact, the ultimate impact of all this alien stereotyping was to give the impression of a cosmos full of alien regimes whose sole purpose seemed to be an irrational plan to invade a relatively lowly Earth.

So, over time, the picture of the imagined alien had begun to degrade. *Doctor Who* began at a time when scientists started to seriously discuss the idea of communication with alien species. And yet too many *Doctor Who* stories featured bug-eyed monsters traveling from the ends of our Galaxy, seeking earthlings to swat. It doesn't matter how you dress it up. They still look like Xeroxed copies of prototyped alien civilizations, projected into the future from our old days of imperialism. Some stories had aliens as colonialist adventurers. Some saw plundering privateers on the make. Others still reimagined plotting conquistadors out for imperial invasion. These fairy-tales seemed totally uninformed by the latest findings of science. They conjure up aliens who seem to occupy a counterfeit cosmos, one bereft of difference in space-time and evolution, where all science lessons were dropped in favor of old-fashioned geocentric nonsense. So sit back and enjoy the *Doctor Who* aliens, monsters, and artificial beings that should never have made the science cut. And let Zygons be Zygons.

Monoids

H. G. Wells would never have put up with this: imagine guys dressed in ill-fitting wet suits, with just the one eye in the middle of their heads, and your grandma's worst wig plonked on top. Exaggerating? If you don't believe me, check out the Monoids on YouTube. Okay sure, for most of its fifty-odd-year run, *Doctor Who* has had to struggle on a limited budget. But that doesn't excuse the kind of poor creative choices never more clearly shown than with these First Doctor humanoids.

Krotons

The Krotons were robots that made an appearance in a Second Doctor story. You simply must look them up on YouTube to believe how stunningly bad they were. Not only did they look like something your teachers would make up for the end-of-year school play, but they also sounded like someone from Monty Python, speaking through a kazoo. This should never have been allowed to happen. Where was their science consultant?

Yeti

More YouTube homework for you. Look up the robotic Yeti, another Second Doctor being, apparently created by a "Great Intelligence." (The very appearance of the Yeti makes the viewer immediately doubt this claim of intelligence.) Something with great intelligence would surely have come up with something more convincing than a gargantuan teddy bear with fluorescent eyes. Great intelligence doesn't come anywhere near it. Another science fail.

Alpha Centauri

I've written elsewhere about the lack of common sense in the naming of some stars and planets in *Star Wars* (for example, the moon of Endor is in orbit around a gas giant named Endor, both of which revolve around the twin stars of Endor and Endor). *Doctor Who* goes a few rungs further up the ladder of stupidity with, get this: an ambassador from Alpha Centauri called Alpha Centauri. How does that even make any sense? Imagine an ambassador from Earth called Earth meeting an ambassador from Mars called Mars. It reminds me also of Trump referring to Apple's Tim Cook as "Tim Apple." All these details are too confusing for some. Anyhow, if the name wasn't bad enough, the alien species, the Alpha Centaurian, was totally scientifically ridiculous. Have you seen those children's drawings of a green alien that is just a huge eye on legs? That's Alpha Centauri (YouTube it. Dare you.)

Giant Maggots

Yeah well, if *Star Wars* can sport asteroidal slugs, why can't *Doctor Who* have Giant Maggots? Not the kind you'd take fishing, you understand. That would be far too scientifically prosaic. No, these fellas are far too large for angling, unless you want to catch Jabba the Hutt. The Third Doctor struggled with these fearsome terrestrial Maggots. If you take a look at the YouTube clips, you'll see they're just ordinary maggots, made to look giant by the judicious use of cunning camera angles. The modern eye is not so easily deceived.

Wirrn

Now insect aliens have made many appearances in the misty history of sci-fi. The one movie that pretty much established the rules of this game was the 1954 film *Them!* Atomic tests in the New Mexico desert create a by-product of giant mutant ants. *Them!* was one of the first-ever movies to have televised advertisement saturation. (Sadly, a proposed late-1950s sequel to be called *It! Son of Them!* never materialized—it would have been one of the all-time greatest movie titles.) But what *did* materialize, unfortunately, was the Wirrn in *Doctor Who*. But, budget lacking, these insects sported the kind of costume your ma might have knocked up for trick or treating. Guaranteed it's going to look lame on TV. Still, it seems the Wirrn were scary enough for the Fourth Doctor, who has some fun running away from the jiggling creatures, as they wander the streets in search of Halloween, perhaps. Admittedly, however, the main criticism here is budget-constrained production values, though more creative science might have made the difference.

Rubber Dinosaurs

Now most of us know the Twelfth Doctor chewed the fat with a realistic and CGI'd Tyrannosaurus Rex in late Victorian London. But, before the days of *Jurassic Park*, it seems the Third Doctor and Sarah Jane Smith had to put up with Barney the Dinosaur's long-lost cousin. About as frightening as your grandma without her teeth in. Actually, *less* frightening than that. This should never have made the science cut. Guys, humans have never coexisted with dinosaurs. We set *that* record straight when Raquel Welch appeared alongside the terrible lizards in *One Million Years BC*—the actual time gap between humans and dinosaurs is around sixty-six million years.

Kandyman

And just when you thought the army of *Doctor Who*'s imagined aliens simply couldn't get any worse, up pops the Seventh Doctor with an adversary entirely made of candy. Yep, you heard me right, candy. Duh; that's why he's called Kandyman. Looking like a basketball player made out of liquorice of all sorts, he just isn't credible. Not even when you're three years of age. All you want to do is eat his face. Come to think of it, this is why Kandyman would have met with an evolutionary cul-de-sac. He'd be far too tasty for prey and competitors alike. Literally eaten out of existence.

Myrka

Myrka was a creature from the inkiest, murkiest depths. It electrifies by touch and is practically indestructible, except for a weakness to ultraviolet light. On paper the Myrka sounds like a real terror. Until, that is, we actually *see* it. On screen, it's got the grace of a Walrus on a beach. Its perambulation resembles two guys lolloping around in a rubber amphibian suit (which is probably what's happening inside the Myrka). You know, like a pantomime horse, but green and camel-sized, with the head of a frog and flippers for feet. Then again, you might argue, what about the duck-billed platypus? When the Australian platypus was first reported by Europeans in 1798, British scientists' initial reaction was that the creature's alleged traits were a hoax.

The Body of Morbius

We've all heard of Frankenstein's creature. Mary Shelley's classic science fiction tale from 1818 was all about a creature made from spare body parts taken from charnel houses. You're already halfway to knowing the Body of Morbius too. Using body parts gathered from various unfortunate aliens, this guy Morbius was built by the scientist Mehendri Solon to house the brain of a most despicable Time Lord, the said Morbius. The result is a mismatched fur-covered beast with one huge lobster-like claw, and a humanoid hand on the other arm. But its head is the icing on the cake: looking like a large toy hamster ball with brass showerheads for antennae. Honestly. I'm not kidding. Go check on YouTube. Again.

How Does *Doctor Who* Use the Science of Exoplanets?

"I know that David Tennant's *Hamlet* isn't till July. And lots of people are going to be doing Doctor Who in Hamlet jokes, so this is just me getting it out of the way early, to avoid the rush . . . "To be, or not to be, that is the question. Weeelll . . . More of *a* question really. Not *the* question. Because, well, I mean, there are billions and billions of questions out there, and well, when I say billions, I mean, when you add in the answers, not just the questions, weeelll, you're looking at numbers that are positively astronomical and . . . for that matter the other question is what you lot are doing on this planet in the first place, and er, did anyone try just pushing this little red button?"

—Neil Gaiman interview (2008)

The Principle of Mediocrity

Doctor Who's Whoniverse relies a lot on a little-known idea in science: the principle of mediocrity. This principle says that, given there's life on Earth, life should also exist on Earth-like planets out in deep space. The principle can also be used to suggest that there's nothing very unusual about our solar system, Earth's planetary history, or human evolution, though the last of these is arguable. Keeping this in mind, you can see that *Doctor Who* is a statement about the place of us humans in the Universe. *Doctor Who* assumes mediocrity in its portrayal of our Galaxy. It's full of alien planets, and alien life forms, whose tech is more advanced and whose powers are greater than ours. *Doctor Who* simply *doesn't* say that humans are special, privileged, and certainly not superior! In short, *Doctor Who* has dreamt up many other worlds.

Another simple statement of *Doctor Who*'s position on planets is this: if the Earth is a planet, then the planets could be Earths. Today, we live in a great age of discovery. We live at a time about which many writers and thinkers have only dreamed. Since the Greek atomists, thinkers have imagined there might be innumerable planets in the cosmos. And yet the first confirmation of an exoplanet orbiting an ordinary star was only made in 1995—that's thirty-two years after *Doctor Who* began. And during those thirty-two years, *Doctor Who* broadcast over a quarter of a century of stories on other planets. Only since 1995 have astronomers discovered such planets. The prediction for the future is stunning: there may be tens of billions of planets in our Galaxy alone. In a very real sense, the Whoniverse has been reborn in reality.

So what are the most notable exoplanets from *Doctor Who* history? And how have the writers evolved their planetary ideas since the very start?

Gallifrey

In the early days of the Second Doctor, the home of the Time Lords, Gallifrey (which remained nameless for a few seasons), was represented by nothing more than a droning sound. The sound pulsed in waves to give the effect of a large planetary expanse. For once, budgetary constraints appear to have delivered here a more creative example of dealing with the science fiction under duress. By the 1970s, Gallifrey had changed again. Now it was pictured full of outdated computer tech, surrounded by spray-painted walls. In more recent series still, we spy Gallifrey as a gleaming citadel, under a burnt-orange sky, the very picture of a blissful utopia.

The Library

In the famous sci-fi movie *Solaris*, there's an alien planet that is entirely sea. The sea of Solaris itself is a sentient, thinking being which encompasses the entire world. In the Tenth Doctor story

"Silence in the Library," we meet a different type of planetary intelligence—an entire planet that is a library. The very idea suggests a populated space so sophisticated that they can assign a planet's resources in this kind of way. It's like the great Library of Alexandria, only in deep space. Indeed, this planetary library is built during the fiftieth century, and contains every book ever written, presumably including this one. All seems fine until the Vashta Nerada hatch from the books . . .

House

The movie *Solaris* may have been an influence on our next Whoniverse exoplanet choice too. The exoplanet House is also a planetary intelligence—and it has a voice! Actually, House is more like an asteroid than a planet, but still proves to be one of the Doctor's best enemies. The landscape is presented as a junkyard, with the junkyard beings of Auntie and Uncle, along with cluttered corridors that hide all sorts of secrets. But *Doctor Who* is also making a serious point: the asteroids have long been suggested habitat for human colonization.

Other Earth

In the history of *Doctor Who* there have also been plenty of explorations of the idea of exoplanet Earths. Other Earth is in one such story. Sure, it may be our own planet, but as it's in a parallel Universe, it's technically an alien exoplanet! This Tenth Doctor story, "Rise of the Cybermen"/"The Age of Steel," features a parallel Earth, one in which zeppelins hover above London, and the population is under threat from John Lumic, owner of a certain Cybus Industries.

New Earth

New Earth is another exoplanet Earth. It features in the Tenth Doctor episodes "The End of the World"/"New Earth"/"Gridlock." After the end of the world, humans find another home. We see two sides to the futuristic New Earth. We see the Doctor and Rose materialize outside a gleaming hospital that fills you with awe and wonder. But in Gridlock, we also see that parts of the planet have become deadly traps.

Krop Tor

Exoplanets in deep space are to be found in orbit about exotic stars, as well as normal stars like our Sun. Krop Tor is such a world. The exoplanet appears in the Tenth Doctor episodes "The Impossible Planet"/"The Satan Pit." News here is that not only is this the first story in which the Ood appear, but it also features a monster who claims to be Satan. And, just in case that wasn't enough for viewers, the Tenth Doctor and Rose become stranded on this planet, which orbits a black hole.

Midnight

Sometimes *Doctor Who* features planets that appear more impossible than they actually are. Take Midnight, for example. Midnight is an airless planet in the system of Xion. The planet is made mostly of diamond glaciers and mountains, which unsurprisingly humans have colonized! Naturally, many exoplanets have their downsides. And here we have a hair-raising Midnight Creature, which lurks in the shadows of this planet where even the rays of sunlight are deadly. But, diamond planet, anyone? Turns out, it's true. For example, 55 Cancri e is an exoplanet in orbit about the Sun-like host star 55 Cancri A. In October 2012, it was announced that 55 Cancri e could be a diamond planet. Huge parts of the planet's mass would be carbon, much of which may be in the form of diamond as a result of the temperatures and pressures in the planet's interior.

Solos

On occasion, *Doctor Who* also experimented with exoplanets that had eccentric orbits. The misty planet of Solos, for example, had a two-thousand-year orbit that brought it in close, and then took it far from its star, resulting in seasons lasting five hundred years. Not terribly well thought-out, as one would imagine conditions on the planet would be unlivable for centuries!

Skaro

And finally to Skaro, the first-ever alien world visited in *Doctor Who*. Skaro is where we first meet the Daleks. They share the planet with the humanoid Thals, a race of fair-haired warriors who are in constant conflict with the more Nazi-minded Daleks. Over time, Skaro was portrayed as a steamy post-apocalyptic world, densely populated by dilapidated buildings—one of which is a huge skyscraper-sized Dalek structure. Among Skaro's few continents lies an "island of gushing gold," where "jets of molten gold shoot into the air." Exactly how this gold got there, and in such cosmic quantities, just happens to slip the writers' minds!

Questions for the Doctor: What's It Like Waking Up on a Space Station?

"The International Space Station is a phenomenal laboratory, an unparalleled test bed for new invention and discovery. Yet I often thought, while silently gazing out the window at Earth, that the actual legacy of humanity's attempts to step into space will be a better understanding of our current planet and how to take care of it. It is not a perfect world, but it is ours. Sometimes you have to leave home to truly see it."

—Chris Hadfield, *Wired* (2013)

The Wheel in Space

Over the years, *Doctor Who* has featured over one hundred space stations. They range in number from the days of the Second Doctor to Platform One in the Ninth Doctor episode "The End of the World." In 1968, the Second Doctor story "The Wheel in Space" was broadcast. The Wheel was an Earth space station, peeking out at the deep Whoniverse, and manned by a small international crew, like today's International Space Station (ISS). The Wheel also acted as a stopping point for deep space ships, a research station, and an early warning monitor for dangerous Sun activity. The Wheel was even armed with an X-ray laser and a force field; no expense spared in the imagination!

In 1975, the year that "The Ark in Space" was broadcast, a group of real-life Earth professors met for ten weeks to design space colonies. They recommended a colony that would have a wheel-like habitat one mile in diameter. Sound familiar? Space colonists would live in a tubular wheel structure, which would rotate to mimic gravity, and use mirrors to draw down sun power.

The idea of space stations has a surprisingly long history in fiction and fact. The first space station ever mentioned was the result of an accident. The "brick moon," conceived by Edward Everett Hale in his 1869 novel of the same name, was supposed to be a satellite, which served as a navigational aid for travelers. Instead, it accidentally launched with passengers aboard.

Brick Moon was followed by other science fiction stories. Most notably, these included tales by Konstantin Tsiolkovsky (who also dreamt up the space elevator!). Tsiolkovsky set out to explore the scientific feasibility of space stations in his professional work as a direct consequence of the inspiration provided by fiction. Later still, German rocket scientist Hermann Oberth became the first writer to actually use the term "space station." It appeared in his 1923 work, *The Rocket into Planetary Space*. Oberth used the term to describe the wheel-shaped, spoke-filled habitat that he believed would serve to transit passengers to Mars.

And it was Oberth's pupil, Werner von Braun, who moved to NASA after an early career in Germany and helped popularize the concept. As a consequence, the wheel-and-spoke design made an appearance in the hugely influential 1968 movie *2001: A Space Odyssey*, among others. It is a design which so far has not been made a reality. (When the first space station Salyut 1 was launched in 1971, it and subsequent stations, including Mir and the ISS, followed a more modular pattern.)

Yes, by the 1970s, humans were living successfully in Earth-orbiting space stations. The Russian Salyut was the world's first crewed station. And that other famous Russian station, Mir, broke endurance records in 1994–1995 when cosmonaut Valeri Polyakov spent 437 days aboard, setting the record for the longest time continuously spent in space. The first internationally crewed space station, like The Wheel, was the ISS. The ISS crew first arrived in 2000, and the mission has since proved that humans can live for long periods in low-Earth orbit. As with The Wheel, experimental research has been done on self-sufficiency in space—crucial knowledge for building a future colony on the Moon.

The Doctor in Space

But what about waking up on a space station? That must be some experience. You might imagine your alarm clock, floating ever so slightly above the bedside table, with a cosmic view of the stars through your porthole. Consider Platform One in the Ninth Doctor episode "The End of the World." It's a future space station far more advanced than our ISS, and a lot bigger. Platform One has guest suites and an enormous engine room, as well as docking bays large enough to permit two ships at the same time. It also has a force shield to protect it from the heat of the red giant Sun and the impact of the exploding Earth.

In contrast the ISS is about the size of a football pitch with the living space of a six-bedroom house. It's got two bathrooms, a gym, and a 360-degree bay window. But the ISS is under constant threat from space debris and micro meteors, as well as harmful solar and cosmic radiation. It's been designed to survive collisions with objects the size of peas and smaller. So, in addition to its aluminum hull, some parts have extra shielding made of, would you guess it, plastic. But what space stations would the Doctor rather *not* wake up on?

The Time Lord's Space Station from "The Trial of a Time Lord"

This station was the pinnacle of the Time Lord justice system. Zenobia was a massive orbiting courthouse where the Sixth Doctor faced an inquiry into his actions, led by the sinister Valeyard. Not a space station the Doctor would want to wake up on. It had an ornate courtroom, handy access to the Matrix—the Time Lords' "library" of all knowledge—and temporal tractor beams. Zenobia might sound like a pleasant place to visit, but you wouldn't want to find yourself on trial there—as one character observed, the Inquisitor seems to be carved out of something hard and nasty.

The Crucible from "Journey's End"

The Crucible was a planet-sized Dalek space station concealed within the Medusa Cascade, the Dalek mothership at the heart of their scheme to destroy the whole of the Whoniverse with a reality bomb. Not a space station the Doctor would want to wake up on, the Crucible was infested with Daleks, including the oversized Dalek Supreme with its red livery and gold adornments. The station also contained a Z-Neutrino energy core capable of destroying a TARDIS! In short, you're most unlikely to make it out alive, unless you're the Doctor.

Space Station Nerva from "The Ark in Space"

The TARDIS brought the Fourth Doctor to Space Station Nerva in the far future. While wild solar winds had ravaged planet Earth, the future of humanity had been put on ice. Unfortunately, the sleepers remained suspended far past their wake-up call due to an infestation of giant space insects called the Wirrn. Not a space station the Doctor would want to wake up on!

Are There Baby Universes Too?

In the Fourth Doctor story "Full Circle" (1980), the Doctor discovers the life cycle of three closely related species on the planet Alzarius—the humanoid Alzarians, the Marshmen, and the Marshspiders. "Full Circle" was the first of three loosely connected stories set in another Universe to the Doctor's own known as E-Space.

Forget the iPhone, Welcome to E-Space

Are there baby Universes out in the deep space of our Universe? There certainly is in the Whoniverse. In the Fourth Doctor story "Full Circle," the Doctor was on his merry way to Gallifrey, as they say. But a wormhole swallows up the TARDIS (what's new?) and the Doctor finds himself on the planet Alzarius. The planet Alzarius has the same position as Gallifrey, but in another (baby) Universe, known as E-Space. The Doctor's own, much larger, Whoniverse being known as N-Space.

To be clear, E-space here stands for exo-space. And N-space, unsurprisingly, stands for normal space. If it helps, you can think of E-space as a more cosmic version of exoplanets—planets outside our Solar System. Just decades ago the idea of exoplanets was merely theory. But telescope technology has since delivered the evidence that there are probably billions of exoplanets out in deep space. The same can't be said for detecting an exo-space. Yet.

Is the Whoniverse anything like our own Universe in this arrangement? Possibly. As you probably know as a Whovian, our Universe is big. Really big. In the words of the Eleventh Doctor, "This is one corner . . . of one country, in one continent, on one planet that's a corner of a Galaxy." And as Monty Python sang about our Galaxy in their "Galaxy Song":

Our galaxy itself contains a hundred billion stars/It's a hundred thousand light-years side to side/It bulges in the middle sixteen thousand light-years thick/But out by us it's just three thousand light-years wide/We're thirty thousand light-years from Galactic Central Point/We go 'round every two hundred million years/And our galaxy itself is one of millions of billions/In this amazing and expanding Universe.

As you know, talk of all these light-years is due to the fact that astronomers measure things in light-years. Light is the fastest thing there is. Light takes over two million years to reach our telescopes from the next galaxy, Andromeda. Andromeda is the closest galaxy to Earth, out of the two trillion other galaxies out there in the observable Universe. In fact, the farthest we can see from Earth is about fourteen billion light-years away, which means it would take fourteen billion light-years to make the journey from the edge of the Universe to Earth.

The story of the Universe began around fourteen billion years ago. According to current theories that are subject to change, a bubble of space and time then popped into reality and began to expand. Before then, there was no space, no stuff, and no time. It could be that as it was growing and evolving over the last fourteen billion years or so, our Universe spawned a number of bubble-like, baby Universes. The question is: If these baby Universes exist, would there be a way of getting from our Universe to the baby bubble Universes? Possibly!

It could be that wormholes would link the main Universe to the bubble babies, so a space traveler could jump from the Universe to a baby, and even from baby to baby. All this means that the idea of E-Space was a very cunning creation by *Doctor Who*. Indeed, the storyline in *Doctor Who* is that the entity that allows travel between the two Universes is known as a Charged Vacuum Emboitment (CVE). The tale told is that a CVE was created by the mathematical brains of inhabitants of the planet Logopolis. They used it to stop the Whoniverse collapsing due to a shortage of available energy.

So perhaps in the future, we may meet an advanced alien intelligence who are able to solve the problem of creating wormholes. Sadly, for the time being, as with exoplanets, we'll simply have to wait until our tech is advanced enough before we can discover the possible existence of an E-space.

Is *Doctor Who* Right: Will There Be Space Vacations?

In the Twelfth Doctor story "Mummy on the Orient Express" (2014), the Doctor takes Clara for one "last hurrah"—on a vacation aboard a space-bound train modeled on the famous Orient Express.

"Well, now you've put your finger right on it. In order to have all of these wonderful things in space, we don't have to wait for technology—we've got the technology, and we don't have to wait for the know-how—we've got that too. All we need is the political go-ahead and the economic willingness to spend the money that is necessary. It is a little frustrating to think that if people concentrate on how much it is going to cost they will realize the great amount of profit they will get for their investment. Although they are reluctant to spend a few billions of dollars to get back an infinite quantity of money, the world doesn't mind spending $400 billion every year on arms and armaments, never getting anything back from it except a chance to commit suicide."

—Isaac Asimov, interview with Phil Konstantin,
Southwest Airlines magazine, 1979

Space Tourism

Space tourism has really taken off in the Whoniverse. We meet passenger liners in the Fourth Doctor tale "Nightmare of Eden," and in the Sixth Doctor story "Terror of the Vervoids." The Fourth Doctor also visits a pleasure complex on the planet Argolis in the story *The Leisure Hive*. More recently, the Tenth Doctor boards a cruise liner (spaceship version) of the RMS *Titanic*, which travels to Earth in December 2008 to allow its passengers to experience a good old terrestrial Christmas. And for her final trip, the Twelfth Doctor takes Clara on board the space-traversing Orient Express, built to travel along "hyperspace ribbons" rather than rail tracks.

And in our own Universe, the Solar System is now open for business. Space tourism will soon be here. Hardly surprising when you think of the science of it. Space tourism begins really close to home. Space is only 62 miles away. In the grand scheme of things, that isn't very far at all. If you live in Seattle, Canberra, Hyderabad, Cairo, Beijing, or central Japan, for example, space would be closer to you than the sea. When trying to get your head around the sheer size of space, and the huge distances we need to cross in the future, it helps to remember that moons go around planets, planets go around stars, stars are arranged into huge galaxies, and galaxies make up the large-scale structure of the Universe.

Space Stations

Science fiction like *Doctor Who* has long dreamed of venturing into space. So far, however, only a few hundred astronauts and cosmonauts have made the trip. That may soon change. Companies are currently planning space tourism projects, as long as you have the money to pay for the flight. And that's been the story in *Doctor Who* too. It's often private companies that have ventured into the exploration of space.

One of the fictional front-runners on the idea of space tourism was British science-fiction writer Arthur C. Clarke. He imagined businesses working on the moon in *A Fall of Moondust* in 1961, along with American film director Stanley Kubrick on the famous 1968 movie *2001: A Space Odyssey*. The movie was one of the very first to carry "product placements," with companies such as Bell, IBM, Pan Am, and AT&T all featuring in the film. It was a prophetic vision of future space travel, complete with corporate logos and trademarks, showing a world absolutely managed by private capital. *A Space Odyssey* also dreamt up space tourists, being served drinks by a stewardess, in a station orbiting the Earth. It all looked so very civilized.

Back in reality, the possibility of space tourism got a lot closer in 2004 with the Ansari X-Prize. This was a competition to design a reusable way of taking people into space, getting them back to Earth, and relaunching again within two weeks. The prize was won. The technology is within reach. And now companies are developing flights into space for the twenty-first century ahead. At the moment, of course, there's hardly anywhere to actually go. At least not until a tourist space station is built, such as those we spoke of in past episodes of *Doctor Who*. For now, passengers will be taken just above the 62 miles barrier that separates Earth from space, and then they will be brought back home. It's much like a helicopter trip around the Statue of Liberty, only darker and more costly.

Companies are now actively dreaming up space stations to place in orbit about the Earth. Bigelow Aerospace is working on the idea of an inflatable space hotel. You can imagine taking a dip in a space station swimming pool, with a glass bottom, and through which you can see the Earth below. Let's hope gravity loss doesn't cause the kind of giant-blob-of-water out-of-pool experience that happened to Jennifer Lawrence in *Passengers*. As we learned from the Second Doctor story "Wheel in Space," orbital stations can be made to spin, which makes a kind of gravity that grounds those inside. Such spins can be spotted not only in *Doctor Who*, but also in the Cloverfield Station in *The Cloverfield Paradox*. Stations would get the same sunlight as the Earth, and can be built with dome sections where food could be grown. These space cities could house hundreds or even thousands of people in the future. Companies are no doubt waging that such space experiences will be easily sold as the holiday of a lifetime. Indeed, space tourism is very popular with the public. A BBC poll found over 70 percent wanted a holiday for one or two weeks in space, 88 percent wanted the excitement of a spacewalk, and 21 percent wanted to chill out in a space station hotel.

Gateway Foundation

In fact, the public may get the chance to chill in space sooner than they expect. An American company has a mission to make space vacations a reality by 2025. The Gateway Foundation is busy building the very first space hotel with artificial gravity. Their aim is that their Von Braun Space Station will be the first commercial space station, according to Tim Alatorre, a senior design architect with the company.

Named after Wernher von Braun, the ex-Nazi aerospace engineer and space architect, the Von Braun Rotating Space Station will be a space hotel for paying clients. The station design works somewhat like the wheel of a bicycle, with spokes coming out of a central hub. Spaceships will dock at the hub, dropping off excited customers who will live on the perimeter, with the station's rotation creating the artificial gravity.

Though artificial gravity sounds like something out of *Doctor Who*, The Gateway Foundation say the science is sound, with company designers using tech from the International Space Station. The Foundation's mission is very dependent on Elon Musk and SpaceX's launch system. SpaceX is creating the Super Heavy and the Starship platform. The expectation is that ticket prices are going to get lower in time. So, whereas prices are now prohibitive for most people, The Foundation hope a space vacation will become a common thing people do in the future. As soon as SpaceX Starship is ready to launch and is orbital, The Foundation want to be one of the first customers to launch into orbit.

So, the plan is that 2025 is the date when people will be able to vacation in space. The Von Braun Space Station is hoped to be a big draw, housing not just a hotel but also a restaurant, bar, and gymnasium. But, significantly, The Foundation hope it'll be the first step to many more humans advancing beyond the atmosphere, even if at first they will be members of a rather elite club. The Foundation hope to build a space industry. They picture multiple stations in space—with their space tours involving trips to the Moon, to Mars, as well as to other space stations. With clients being able to go up into space and look back on the blue marble Earth, they hope their mission is going to have a profound impact on all of us.

Space Elevator

Finally, consider the space elevator. If space tourism and space stations are to become a reality, we'll need a cheap and regular way of getting back and forth into space, right? Some way that doesn't use up too much rocket fuel, and is kind to the environment. That's where the space elevator comes in. Imagine jumping into an elevator and pressing the button marked "space" or even "space station." Such a thing was imagined in an 1895 book called *Daydreams of Heaven and Earth*, by Russian rocket pioneer Konstantin Tsiolkovsky, who wrote science fiction as well as science.

At heart, the design of a space elevator is a 29,000-mile cable, made from an as yet undeveloped material. This "beanstalk" material would need to be thirty times stronger than steel, and have a diameter of no more than four inches. Tricky. This super-strength beanstalk stuff would be tethered to the space station to enable continuous transport into space. NASA has spent millions of dollars running a competition to design space elevators. To date, the space elevator idea has failed to capture the public imagination. And yet the world needs a space elevator. Our future in space demands the cosmic power behind an elevator button marked "space."

What If *Doctor Who* Stories Fell like Cinema Rain?

Imagine you're a forensic scientist. Kind of. Your job is not to collect, preserve, and analyze scientific evidence, but to collect, preserve, and analyze the very best of *Doctor Who*. Today, the task with which you've been charged is to analyze ten of the most quintessential tales of the Doctor. Sure, they are mere stories. But remember that famous quote from *Doctor Who* writer Stephen Moffat on the demise of the Eleventh Doctor: "I'll be a story in your head. That's okay. We're all stories in the end."

The theater of your forensic operation is not a lab. Your theater is an ordinary cinema. There are almost a thousand episodes of *Doctor Who*. Imagine they slowly fall now, like silver raindrops from the ceiling of the cinema. Each raindrop contains an episode. An encapsulated story from the time traveler's past and future, frozen inside like some kind of snow globe. One drop, one episode. Which few stories do you collect and preserve? Which best capture the imagination of the longest-running sci-fi TV show in the world, and the most successful sci-fi series of all time?

"The Daleks" (1963): First Doctor

When the robotically evil Daleks were beamed into the second serial of *Doctor Who*, the show skyrocketed in popularity overnight. This episode takes place on the planet Skaro where there are two alien races—the Thals, a race of peaceful humanoids, and the infamous Daleks, malicious mutants encased in robotic shells. If you've ever wondered what a Dalek city might look like, check out this tale.

Choice quotes:

Alydon: "But why [would the Daleks] destroy without any apparent thought or reason? That's what I don't understand."

Ian: "Oh, there's a reason. Explanation might be better. It's stupid and ridiculous but it's the only one that fits."

Alydon: "What?"

Ian: "A dislike for the unlike."

"Blink" (2007): Tenth Doctor

With its themes of time travel, cryptic DVD Easter eggs, and the famous Weeping Angels, this very quotable episode is a favorite of many. It's the perfect introduction for a *Doctor Who* novice as the writing of Steven Moffat here is thrilling. Curiously, the Doctor hardly appears, as the story revolves around a Sally Sparrow, and the strange happenings at a house on the edge of town . . .

Choice quotes:

The Doctor: "People assume that time is a strict progression of cause to effect, but *actually* from a non-linear, non-subjective viewpoint—it's more like a big ball of wibbly wobbly . . . time-y wimey. . . stuff."

The Doctor: "Don't blink. Blink and you're dead. They are fast. Faster than you can believe. Don't turn your back. Don't look away. And don't blink. Good luck."

"Genesis of the Daleks" (1975): Fourth Doctor

As the Daleks had become the best-loved baddies, by the time the Fourth Doctor popped up, a whole ten episodes had been dedicated to them. In this story, Dalek writer and creator Terry Nation explained the origin of the dreaded Daleks. It's a terrific tale of the battle between the Kaleds and the Thals. The battle comes to a climax when Kaled scientist Davros betrays his people and gives the Thals a formula to destroy the dome protecting the Kaleds.

But hold! Davros had been secretly building an army of travel machines that contained the mutant Kaleds—a.k.a. the Daleks.

Choice quote:
Daleks: "We are entombed, but we live on. This is only the beginning. We will prepare. We will grow stronger. When the time is right, we will emerge and take our rightful place as the supreme power of the UNIVERSE!"

"Human Nature"/"Family of Blood" (2007): Tenth Doctor

On the run from the Family of Blood, the Tenth Doctor morphs into a human to hide out. He becomes John Smith, a teacher at a posh school in England, just before World War I. Martha Jones watches over him, as his transformation has robbed him of his Time Lord memories. But the Family of Blood hunts him down and ultimately [spoiler!] the Doctor retransforms to defeat the Family with ruthful vengeance. The Doctor and Martha wonder if anyone would have died had they not chosen that school as a place to hide. It's a question about the Doctor that has popped up before: Does he save people in need, or does his appearance guarantee death and destruction?

Choice quote:
John Smith: "Mankind doesn't need warfare and bloodshed to prove itself. Everyday life can provide honor and valor. Let's hope that from now on this country can find its heroes in smaller places. In the most ordinary of deeds."

"The Empty Child"/"The Doctor Dances" (2005): Ninth Doctor

When the Ninth Doctor and Rose are chased through the London blitz by a strange kid in a gas mask, it's a contender for the most scary *Doctor Who* episode of all time. The story has a totally unexpected happy ending, and also introduces the character of Jack Harkness.

Choice quote:
The Empty Child: "Are you my mummy?"

"The Doctor's Wife" (2011): Eleventh Doctor

The Doctor's Wife is Neil Gaiman's wonderful tale of the soul of the TARDIS shifted into the body of a woman called Idris. The soul shift is at the bidding of a being called House who devours Time Lords. The episode alters the way we think of the Doctor's backstory; Doctor or TARDIS, who stole whom?

Choice quotes:
The House: "Fear me. I've killed hundreds of Time Lords."
The Doctor: "Fear me. I've killed all of them."

"The Caves of Androzani" (1984): Fifth Doctor

An underground *Phantom of the Opera* on the planet Androzani Minor, this episode is an epic finale for the Fifth Doctor, who sacrifices his life in the most heroic fashion. The scar-faced, drug-dealing masked villain, Sharaz Jek, is one of the show's most developed and complex characters.

Choice quotes:
The Doctor: "Androzani Major was becoming quite developed last time I passed this way."
Peri: "When was that?"
The Doctor: " . . . I don't remember. I'm pretty sure it wasn't the future."

"A Good Man Goes to War" (2011): Eleventh Doctor

In order to find and rescue Amy Pond, the Doctor calls in a few debts [from some of those he has helped] across time and space. This, the 777th episode of *Doctor Who*, neatly ties up the mystery of River Song and we also get a glimpse at the Doctor's own baby cot. It's also where we first meet Madame Vastra, Jenny Flint, and Strax, as well as the nonliving Headless Monks. Having no heads the monks can never be dissuaded, afraid, or surprised.

Choice quotes:
Rory: "I have a message and a question. A message from the Doctor and a question from me. Where is my wife . . . ?"
Cybermen: "What is the Doctor's message?"
Rory: [As a whole Cybermen fleet suddenly explodes] "Would you like me to repeat the question?"

"Robots of Death" [1977]: Fourth Doctor

Doctor Who expert Peter Grehan says this, his favorite episode, has lots of references to robot folklore. For example, the villain Taran Capel is a reference to Karel Capek, the man who wrote the first-ever robot story. The story also shows the importance of human body language and the effect this might have on humans working with robots that don't have it. At the time of filming, recent research had shown that women were better body language readers (as demonstrated by Leela).

Choice quote:
The Doctor: "You're a classic example of the inverse ratio between the size of the mouth and the size of the brain."

"Father's Day" [2005]: Ninth Doctor

Rose and the Ninth Doctor travel back in time to the day when Rose's father died when she was a small child. The Doctor's not happy about the trip, so he tells Rose not to interfere. This ground-breaking episode showed an emotional connection between a time-travel companion and the Doctor, unlike most of the classic series of old.

Choice quotes:

Young Mickey: [running to the church] "Monsters! Going to eat us!"

Suzie: "What sort of monsters, sweetheart? Is it aliens?"

Part II
Time

Introduction

"Absolute, true, and mathematical time, of itself, and from its own nature, flows equably without regard to anything external, and by another name is called duration: relative, apparent, and common time is some sensible and external (whether accurate or unequable) measure of duration by means of motion, which is commonly used instead of true time; such as an hour, a day, a month, or a year."

—Isaac Newton, *Principia* (1687), Andrew Motte Tr.
The Mathematical Principals of Natural Philosophy (1803)

"Time as a fourth dimension rests vertically on the other three—just as in space the vertical juts out of the two-dimensional plane as a third dimension. Distances through space-time comprise four dimensions, just as space has three. The more you go in one direction, the less is left for the others. When a rigid body is at rest and does not move in any of the three dimensions, all of its motion takes place on the time axis. It simply grows older."

—Jürgen Neffe, *Einstein: A Biography* (1956)

Doctor Who is, naturally, obsessed with time. The program was originally intended to be an educational series. Two of the First Doctor's companions were teachers. And the TARDIS was meant to take the form of an object from that particular episode's time period; an incognito column in Ancient Greece, perhaps, or an unsung sarcophagus in Egypt, and so on. But a deep time fever soon caught hold. *Doctor Who* began to wonder what might happen if we could tamper with time. Or if we could jump timelines, guide evolution, rewrite history, or even cheat death (which, of course, the Doctor does through regeneration).

Earlier on, we talked about the nature of the science fiction of *Doctor Who*. About the way in which sci-fi looks at the relationship between us humans and the nonhuman but natural world around us. The world revealed by science. Since the scientific revolution half a millennium ago, science has come to dominate all walks of life. Science has sought not just to explore, but to exploit nature. To master it.

Starting in the seventeenth century, it began to dawn on scientists that time was limitless, and inhumanly vast in scale. And by the time of the industrial revolution, monstrous machines were turning over the soil of the world. Dinosaurs were discovered. And the death toll of extinction rang for the first time. The first fossil record was found. It seemed to spew out evidence of creatures no longer to be found on the planet. The new theory of evolution forced us all to confront the terrible extent of history. What if we humans also became extinct? Little wonder *Doctor Who* portrays a planet constantly in peril. But the planet is quite safe; it's future humans that face the real danger.

Suddenly, science was faced with the ultimate challenge. What if we could master time? That brutal agent that devours life and beauty. So began a science fictional obsession in which *Doctor Who* is truly schooled. Sure, there had been flirtations with time before sci-fi. Folkloric time-slip romances where magic is mashed up with myth, and time is lost as a convenience in the story. But the modern idea of mechanized time travel appeared with industrialization. The invention of time travel was caught up with the concept of time itself. The ancient Greeks had two words for time, *chronos* and *kairos*. *Kairos* suggested a qualitative type of time, moments in which something special happens. *Chronos* was a more quantitative time, concerned with measured, sequential events. The industrial revolution brought a machine approach to nature. *Chronos* became king.

And so time travel was born. It was English writer H. G. Wells who gave science fiction one of its most enduring devices. The time machine. With Wells the TARDIS was essentially born too. As Peter Grehan points out in *Connecting Who: Artificial Beings*:

> We can already see elements of Terry Nation's first Dalek story falling into place: the evil technological Daleks, living deep beneath their abandoned city, are the equivalent to the subterranean, machine-obsessed Morlocks, while the peaceful Thals stand in for the Eloi who, although diminutively childlike in the novel, were very Thal-like, tall, handsome and blonde, in their depiction of the 1960 film adaptation. As well as the Morlocks, aspects of the Daleks seem to have been further influenced by the antagonists of another of Wells's works, *The War of the Worlds*. This is an alien invasion story, but the important thing about it for Nation was the "design" of the alien Martians: they were cyborgs.

The long road to the Doctor and almost a thousand time machine episodes starts with Wells. His *Time Machine* is an ingenious voyage of discovery. And Wells's Time Traveler follows a fate like the Doctor's. Both set out to marshal and master time. But both discover the inevitable truth; time is lord of all. The significance of the phrase "time machine" takes on a new meaning. We are trapped by the mechanism of time. Bound by a history that likely leads to extinction. The very stars grow old and die.

Nonetheless, the Doctor continues in his quest to tweak time. To make the future a brighter place, keep us safe from Daleks. The Daleks are a brutal force of evolution. Could they be the "men" of the future? They are alien, yet they were humanoid. They are what we may one day become, with their overdeveloped brains and emaciated bodies. They are the tyranny of intellect without empathy and love.

As *Doctor Who* has shown so often, time has other tricks up its sleeve. Evolution, for example, is a process that reveals itself in time, and so tales often hinge on the question of what might happen in the human future. And time travel allows the Doctor to jump freely between alternate timelines, alternate Earths. Each timeline has its own plausible future. In this way, *Doctor Who* has greatly influenced our culture, inspiring questions such as: How open is the future? Do we really have free will? Isn't all history in a sense a fiction? How can we ever know anything about time other than the fables we create? After all, we are all stories in the end.

So turn into the pages of time, where evolution becomes melodrama. Let's follow the worm-holing TARDIS and plunge into a Time Vortex of our own making, where the Whoniverse conjures parallel worlds, faster-than-light speeds, and the Blinovitch Limitation Effect!

Questions for the Doctor: How Do You Travel through Time?

In the Tenth Doctor story "Blink" (2007), the Doctor is trapped in 1969 and tries to communicate with a young woman in 2007, Sally Sparrow, to stop the statue-like Weeping Angels from taking control of the TARDIS. The story also feats the Doctor's now-famous description of time as a big ball of wibbly wobbly . . . timey-wimey . . . stuff.

"There have been mountains of nonsense written about traveling in time, just as previously there were about astronautics—you know, how some scientist, with the backing of a wealthy businessman, goes off in a corner and slaps together a rocket, which the two of them—and in the company of their lady friends, yet—then take to the far end of the Galaxy. Chrono motion, no less than Astronautics, is a colossal enterprise, requiring tremendous investments, expenditures, planning . . ."

—Stanislaw Lem, interview, *The Star Diaries* (1957)

Time Travel

You know how time travel works in *Doctor Who*. Someone is minding their own business when, before they know it, their time-stream is slipping, there's a weird sound like a key scraping along piano wire, and a blue 1960s-style London police box suddenly materializes in their living room. It seems so easy, if a little surreal.

Time travel must surely be a long-held human aspiration, at least in spirit. The desire to go back and change silly mistakes in life. The nostalgia of wishing to relive heartfelt memories. Or the intellectual curiosity for visiting the past. As Marcel Proust once said:

> Thanks to art, instead of seeing one world only, our own, we see that world multiply itself and we have at our disposal as many worlds as there are original artists, worlds more different one from the other than those which revolve in infinite space, worlds which, centuries after the extinction of the fire from which their light first emanated, whether it is called Rembrandt or Vermeer, send us still each one its special radiance.

Indeed, stories proliferate in which whimsical time travel is a major part of the plot. Think of the temporal hijinks in Charles Dickens's *A Christmas Carol* (1843), or the burlesque and ludicrous book that is Mark Twain's *A Connecticut Yankee in King Arthur's Court* (1889). But such adventures start with a blow to the head, or dreamy magic, before the time travel begins.

That's why the 1895 H. G. Wells novel *The Time Machine* is so important to *Doctor Who*. With Wells the modern idea of time travel begins. With this story, Wells was the first to really tamper with time. His journey at least had a scientific veneer. Wells suggested that, if space is made up of three dimensions, then time must be the fourth. And, if you can move freely in the first three dimensions of space, why shouldn't you also do so in time?

Scientifically Speaking

Wells was aware, as all good Whovians are, that we travel in time every day. Sure, it's only at the rate of one second per second. This, allegedly, is normal. If you happen to have a telescope handy, then stargazing is another form of time travel. The Universe is so vast that starlight from the outer reaches of the Universe takes longer than twice the age of the Earth to reach our telescopes. One of the

brightest stars in the sky, Sirius (after which J. K. Rowling named the character Sirius Black), is about eight light-years away (that's over 47 trillion miles). Light, the fastest thing there is, takes more than eight years to get to Earth from Sirius. We see the stars as they were in the past.

Modern notions of time travel, with all its complexities, date back to 1905. That's the year in which famous Nobel Prize–winning scientist Albert Einstein came up with his special theory of relativity. The theory suggested that time and space are closely linked. His general theory of 1916 said that space and time are pliable. In Einstein's Universe, space-time can be warped, bent, expanded, or contracted in the presence of energy or matter. And *that* means if you fill space with an exotic form of energy, it can be warped so that space and time bend back on themselves, allowing a potential traveler, with the right kind of tech, to tinker with time.

How The Doctor Does It

It seems that time travel in *Doctor Who* can be achieved by a number of means. There's the judicious use of mirrors as in the Second Doctor story "Evil of the Daleks." Some kind of glitch in the fabric of space-time leads to time travel in the Fifth Doctor story "Mawdryn Undead." Or there's even time travel as some kind of natural biological ability, like that of the Tenth Doctor story "Blink," where the Weeping Angels were able to send others back in time with a mere touch. But, most of the time. the Doctor uses a space-time vessel, in the guise of the TARDIS, to time travel at leisure, just like the Traveler in H. G. Wells's original *Time Machine*.

So how scientifically possible is time travel? British physicist professor Stephen Hawking always refused to believe time travel was possible. He often said when asked, "If time travel really *is* possible, then where are the time tourists of the future? Why aren't they visiting us, telling us all about the joys of time travel?" Professor Hawking's argument was taken a little further by author

Brian Clegg in his 2011 book *How to Build a Time Machine: The Real Science of Time Travel*, in which Clegg asks, "If time travel is possible, why not flag a certain place and time in history and invite time travelers to attend? As long as information on the event percolated into the future—and a combination of Internet, print media, and TV coverage would seem to guarantee this unless our civilization were destroyed—how could any time traveler resist?"

What Are the Definitive Time Machines of *Doctor Who*?

"That is the germ of my great discovery. But you are wrong to say that we cannot move about in Time. For instance, if I am recalling an incident very vividly I go back to the instant of its occurrence: I become absent-minded, as you say. I jump back for a moment. Of course we have no means of staying back for any length of Time, any more than a savage or an animal has of staying six feet above the ground. But a civilized man is better off than the savage in this respect. He can go up against gravitation in a balloon, and why should he not hope that ultimately he may be able to stop or accelerate his drift along the Time-Dimension, or even turn about and travel the other way?"

—The Time Traveler in H. G. Wells's *The Time Machine* (1895)

"Long ago I had a vague inkling of a machine . . . that shall travel indifferently in any direction of Space and Time, as the driver determines."

—The Time Traveler in H. G. Wells's *The Time Machine* (1895)

"The Doctor is James Bond but cooler. Bond gets a boat, the Doctor gets a time machine."

—Matt Smith, Eleventh Doctor (2010)

Time Machines

Your time machine is complete. To when in time do you first travel? And what famous events in history make your top three? A short journey back to 1969, perhaps, to follow the Apollo 11 crew as they approach the Moon (keeping a careful distance, of course, as you don't want to end up in Area 51). Or maybe you journey to the prehistoric past? To witness the stepmother of all extinctions unfold firsthand would be something, before beating a hasty retreat in your time machine. Or maybe you travel back in time to 117 AD, when the grand Roman Empire under Emperor Trajan had its greatest reach. Days of the dinosaurs, you say? Set the controls for sixty-five million years ago, or something like that.

Checking time machine systems: Temporal gearbox—check. Fourth-dimensional manipulator—check. Ignition—check. Ignition? Wait, what? Exactly how are these time machines actually meant to work? Let's travel back in time into the history of *Doctor Who* and take a look at the program's record on time machines.

TARDIS

The coolest time machine in the Whoniverse. Also called the Ship, or simply the Box, TARDISes are organic, sentient beings that travel in space-time. (Witness the fact that the TARDIS once told the Eleventh Doctor that while it may not have always taken him where he *wanted* to go, it had always taken him to where he *needed* to go.) TARDISes often mourn the deaths of their Time Lord pilots, some committing suicide by throwing themselves into a nearby star, or flying headlong into the Time Vortex. Legend has it that, somewhere at the end of time, there's an "elephant's graveyard" of TARDISes.

Mode of Time Travel:
Allegedly TARDISes usually moved through space-time by "simply" disappearing from one place and reappearing in another. As the process is known as "materialization," it's hardly surprising that TARDISes were controlled by a component called the dematerialization circuit.

Dalek Time Machine

The Dalek Time Machine, or Dalek time ship, was much like a TARDIS. It was also dimensionally transcendent (bigger on the inside), and was built to chase the Doctor through time and space. But check this out: Dalek time machines also had remote controls. And *that* means you could control them from their point of destination.

Mode of time travel:
Much like a Time Lord TARDIS, the Daleks used the rarest mineral in the Whoniverse, taranium, to power their space-time craft. Time travel doesn't come cheap.

SIDRAT

Another dimensionally transcendent time machine, the SIDRAT stood for Space and Inter-Dimensional Robot All-Purpose Transporter. SIDRATs were used by the War Lords and built with the help of the War Chief, a renegade Time Lord. SIDRATs were less sophisticated than TARDISes, and suffered from their short range and power costs.

Mode of time travel:
Much like a bargain basement TARDIS, built as it was by a Whoniverse version of a second-hand time-machine dealer.

Time Cabinet

This one's a bit of a problem. It looked nice enough, like the kind of lacquered cabinet you might see in your grandma's house. But its engine (which used a zygma beam, apparently) could not only cause huge implosions, but also lead to life-threatening mutations in the traveler. This would be a little like catching a train and finding out you had two heads by the time you got to your destination. No wonder the Fourth Doctor destroyed the key, making the Time Cabinet useless.

Mode of time travel:
Opened by a "time key" in the form of a trionic lattice, a unique element which was central to its operation. Again, not cheap.

TOMTIT

TOMTIT stood for Transmission Of Matter Through Interstitial Time. Possibly the most stupidly named time machine in all of sci-fi. It was originally built at the Newton Institute to break down objects and teleport them. But the Master altered TOMTIT to act as a time-travel device.

Mode of time travel:
The transportation of objects by a teleportation or mass transference.

Osmic Projector

The Osmic Projector is Sontaran tech, which could be used as a time-travel device. Use of the projector would leave a ghostlike figure of the traveler from some minutes after use, which was one of the explanations the Tenth Doctor gave for ghosts. The Sontarans used the projector to invade the human outpost on Agni.

Mode of time travel:
The Fifth Doctor suggested the Osmic Projector system was based on quantum tunneling rather than typical matter transmission, which it sometimes also used.

Time Scoop

A Time Lord device used to remove objects or individuals from one point in space-time to another. The Scoop has been used to transport many of the Doctor's enemies out of harm's way, including a Dalek, a Cyberman, and even a Robot Yeti!

Mode of time travel:
A device that could be targeted at individuals or objects to move them from one point in space-time to another. In short, mass transference.

Circle of Mirrors

The Circle of Mirrors, also called a dimension cannon by Rose, was a time-travel gadget built by the alternate UNIT of Donna's world. It was created using the power of the Tenth Doctor's salvaged TARDIS. But mirrors were the key, as with other time devices used to travel to different dimensions, such as those used by Tharils and Mirrorlings.

Mode of time travel:
How's this for scientific vagueness: the mode of operation was, predictably, a series of mirrors arranged in a circle for unique purposes. This *Doctor Who* time-travel device used 144 separate polished mirrors in a closed space and applied static electricity to achieve a form of projected time travel. In short, the writer wasn't even *trying* to be convincing on this one.

Time Transfer Device

Appearing in the 1972 serial "Day of the Daleks," this portable time machine belonged to guerillas from twenty-second-century Earth. A much cruder forerunner to the Vortex Manipulator, once activated anyone caught in its time field can be transported at least two hundred years through time (the time dial must be broken).

Mode of time travel:
It features a mini-dematerialization circuit and also a time transmitter that allows its coordinates in space-time to be traced from the twenty-second century.

Kartz-Reimer Module

In a region of the Galaxy known as the Third Zone, the scientists Kartz and Reimer have built their own pyramid-shaped version of the TARDIS. But there are snags. It's only big enough to fit one person, and it doesn't work properly due to a safety feature built into the Time Lord technology by Rassilon. As such, the module will only work if it's primed by the "symbiotic nuclei" held within the cell structure of a Time Lord. The module is rigged by the Second Doctor to only work once—it explodes on its next use.

Mode of time travel:
Twisted TARDIS power.

Anyone Heard of the Blinovitch Limitation Time Effect?

In the Fifth Doctor story "Mawdryn Undead" (1983), the story is set in an English boarding school and a spaceship above the Earth in 1977 and 1983. The story also features the Blinovitch Limitation Effect, a fictional principle of time-travel physics in the Whoniverse: first, that a time traveler cannot "redo" an act that he has previously committed, and second, that a dangerous energy discharge will result if two temporal versions of the same person come into contact. The effect was introduced in "Day of the Daleks" (1972).

Time Constraints

What should the Doctor not do? Well, I guess there are a number of answers to this question. But when it comes to time travel, according to the physics of the Whoniverse, one thing he certainly should not do is meet himself. So what does he do? He meets himself. Regularly. The most memorable example of this is during the program's fiftieth anniversary episode, "The Day of the Doctor." The story (which really should be called "The Day of the Doctors") features a meeting not just of the Tenth and Eleventh Doctors, but also the War Doctor (not to mention a very brief glimpse of the then-upcoming Twelfth Doctor, and a guest appearance by the Fourth Doctor; in short, all the best Doctors in one episode).

Professor Einstein's general relativity theory does not prohibit time travel. And a number of physicists in the past have worked out solutions that allow closed time-like loops, which enable repeating space-time coordinates. The most famous of which is the 1988 paper *Wormholes, Time Machines, and the Weak Energy Condition*, by Michael Morris, Ulvi Yurtserver, and 2017 Nobel Prize–winner, Kip Thorne. In short, the paper's message is this: harnessing the energy created by a traversable wormhole could possibly produce a time machine.

Thorne and his colleagues start their paper by looking at any limits the laws of physics place on advanced tech. They conclude that traversable wormholes are technically possible. Under particular conditions, the wormhole could act as a diverging lens enabling travel backward through time. It appears that science and fiction concur when it comes to the challenge of interfering with the past.

In physics, there is a "self-consistency principle" for time travel. It theorizes that any closed time-like curve must be self-consistent, among other things. These laws of physics remain the same for time travelers too. And that limits to zero the probability that any event can cause a paradox, while still permitting free will. In other words, humans have this idea that we can somehow control destiny, and so we behave accordingly. In *Doctor Who*, of course, these themes are explored in a number of ways, as the Doctor and his companions try to use their knowledge of the future to shape events. But there are forces that get in the way of that for most people, and make them realize that control may be an illusion.

The Blinovitch Limitation Effect

Clearly the Doctor isn't most people. Science fact has the "self-consistency principle." The main physical law about time travel in *Doctor Who* is known as the Blinovitch Limitation Effect, or BLE. This law says that if there's a repeated attempt to change history in the Whoniverse, the BLE will kick in, and often with catastrophic results. This means that Time Lords like the Doctor should try to stop repeated attempts to change history so that the BLE doesn't kick in.

Let's try to explain how this law of physics from the Whoniverse acts in a way to get rid of errors and inconsistencies in time.

Imagine a soccer ball goes into a wormhole and back in time. What if the path of the soccer ball is aimed so that when it pops out at the other end of the time loop, it bumps into its earlier self and stops it from going into the wormhole in the first place? Some scientists suggest that this will never happen. And that's because they think the laws of a Universe will automatically correct such errors. The soccer ball would always take a path so that when it emerged at the other end, it would knock its earlier self *into* the wormhole, rather than away from it.

Another Relativistic Issue for The Doctor: The Twin Paradox

There are other puzzles and paradoxes that arise from Einstein's general relativity theory. One is the so-called twin paradox. Imagine there are two Doctors. You can pick whichever is your favorite and simply clone them in your mind. Naming might be a problem, so let's call them Doctor A and Doctor B. After celebrating their twenty-first birthday, Doctor A says *au revoir* to Doctor B, jumps into the TARDIS, and begins his or her journey.

Let's now imagine that the TARDIS comfortably hits a travel speed of the speed of light. Doctor A then travels for seven years toward his or her destination, according to the TARDIS clock. The TARDIS then swiftly slows down, reverses her direction, and travels back to Earth at the same speed as her outward journey. This once more takes seven years of Doctor A's time. So, when Doctor A returns to planet Earth, her age is twenty-one plus fourteen, or thirty-five years old.

Now comes the paradox. When Doctor A greets Doctor B she is amazed to see that Doctor B is seventy-one years old—that's more than twice her age! (Okay, Doctor A may not be *that* amazed as she's the Doctor and knows all about Professor Einstein's general relativity theory, but you get the idea.) What has happened to the Doctors? Simply speaking, the twin Doctors have had different experiences.

A fuller explanation requires math, of course, but it's best explained here by remembering that famous adage of popular relativity theory: *moving clocks run slow*. One of Einstein's great realizations was that time is not absolute. We automatically assume that the time that ticks on the TARDIS is at the same rate as that on Earth. That is not the case. And the slowing down of time on the TARDIS applies to *all* clocks, including the biological clock of Doctor A's body. And a vital point in getting to grips with the paradox is this: the twins don't have identical experiences, only Doctor A feels the deceleration and acceleration when the TARDIS turned around.

Another Time Travel Issue for The Doctor: The Grandfather Paradox

Another science fiction time-travel dilemma is known as the Grandfather Paradox. Basically speaking, it's named after the idea of traveling back in time, accidentally killing your grandfather (one assumes you wouldn't be interested in purposely killing him in some rather complex attempt at delayed suicide), and preventing one of your parents from being born. Which obviously means that you would never have been born in the first place, thus making it impossible for you to have traveled back to accidentally/purposely kill your grandfather. No doubt this is the kind of thing that both the "self-consistency principle" and the Blinovitch Limitation Effect are theorized to prevent!

But the award for the most complex and intriguing science fiction story involving time-travel paradoxes goes to "–All You Zombies–." The story was written by American author Robert A. Heinlein. It was written in one day, July 11, 1958, and published in the March 1959 edition of *Fantasy and Science Fiction* magazine, after being rejected by *Playboy*.

"–All You Zombies–" is the time-travel tale of a young man, who is later revealed to be intersex. He journeys back in time and is tricked into impregnating his younger, female self (before he

underwent sexual reassignment surgery). Thus the tale's protagonist turns out to be the offspring of that union, with the paradoxical result that he is his own mother and father. As the tale unfolds, all the main characters are revealed to be the same person, at different stages of her/his life!

Questions for the Doctor: What Are Your Best Stories of Traveling in Time?

"Time travel is like visiting Paris. You can't just read the guidebook, you've got to throw yourself in. Eat the food, use the wrong verbs, get charged double, and end up kissing complete strangers."
—Ninth Doctor, "The Long Game" (2005)

"If we could travel into the past, it's mind-boggling what would be possible. For one thing, history would become an experimental science, which it certainly isn't today. The possible insights into our own past and nature and origins would be dazzling. For another, we would be facing the deep paradoxes of interfering with the scheme of causality that has led to our own time and ourselves. I have no idea whether it's possible, but it's certainly worth exploring."
—Carl Sagan, interview, *NOVA*, October 12, 1999

Sure, I admit some of you may be thinking: Hang on, aren't all *Doctor Who* tales somehow about time travel? After all, the Doctor *is* a time traveler. But this is a proposed list of the way in which the Doctor might answer the rather long question: "Doctor, what are your best timey-wimey stories, the ones in which you've managed to use the science of time to undo deaths that have been done, to rewrite lives, to spawn parallel worlds and alternate timelines, or even to conjure up a time-traveling wife who keeps on meeting you in the wrong order?"

"Blink" (2007): Tenth Doctor

What better place to start than the story that invented the very phrase "timey-wimey"? This "Doctor-light" story makes an asset of the Doctor's absence and creates a complex time plot, which captured the public's imagination more than any episode since the Daleks first dawned. Arguably. The story features the horrific Weeping Angels who actually feed on time, like many photosynthesizing plants on Earth feed on light. In the (open) eyes of many, "Blink" cannot be beaten.

This story's science innovation:
A team of physicists from Cornell University have now shown that matter can be "quantum locked," or frozen in place. Thankfully, the team's research doesn't prove that Weeping Angels exist. But it *does* show that the idea behind their movement can be demonstrated in the lab. According to one team member:

> In the television show, though, there's something about conscious observation that makes this work; the photons bouncing off the angels have to land in someone's eye to freeze them in place. In reality however (extrapolating generously from this experiment), such a creature could only move in complete darkness, or perhaps only under certain wavelengths of light. For these atoms, it's not the photo that freezes them in place, it's the camera's flash.

"The Pandorica Opens"/"The Big Bang" (2010): Eleventh Doctor

One of the biggest-ever blockbuster finales to a series, "The Pandorica Opens"/"The Big Bang" is timey-wimey in the extreme. Even though we've often been told the Doctor can't meddle in past events, he does so here, and for good reason—the Whoniverse is disintegrating. The Doctor uses one time-travel device (River's vortex manipulator) to free himself and steer the Pandorica into the exploding TARDIS to reboot the Whoniverse. Not only that but the story also featured the largest number of individual alien species ever seen in a single episode.

This story's science innovation:
Not so much an innovation as a playful use of the Second Law of Thermodynamics: the total entropy of an isolated system can never decrease over time. Entropy is a measure of the chaos, or disorder, of a system. And in "The Pandorica Opens" the Whoniverse's disintegration shows the cosmos is winding down in an irreversible entropic decay.

"The Day of the Doctor" (2013): Many Doctors

In "The Day of the Doctor" we meet a War Doctor on the cusp of destroying Gallifrey with the aid of The Moment, a sentient superweapon capable of destroying Gallifrey, Daleks and whole galaxies within a single moment. Even the bomb in this episode is based on time. Versions of the Doctor meet to think through their actions, and come up with an ingenious solution—of shunting their home world sideways in space-time—so that Gallifrey falls no more.

This story's science innovation:
The science of superweapons. Science and sci-fi have been feeding off each other about superweapons for over one hundred years, ever since H. G. Wells literally invented the "atomic bomb" in his 1914 story, *The World Set Free*. The innovation in "The Day of the Doctor" is that the bomb is one of time, rather than space and matter.

"The Girl in the Fireplace" (2006): Tenth Doctor

Time-travel romance. The Doctor discovers time portals on a fifty-first-century spaceship, SS *Madame de Pompadour*. He winds up in eighteenth-century France and meets the lady the ship has been named for, the famous Madame de Pompadour, a.k.a. Reinette Poisson. He keeps popping up in Reinette's life, only minutes apart from him, and each time the pair grow closer. But the time portals exist so that the ship's clockwork droids can get their hands on Reinette's brain. The Doctor defeats the droids and promises to come back for Reinette and show her the stars. Alas, when he returns, Reinette is no more, and the Doctor is gutted.

This story's science innovation:
As Peter Grehan points out in *Connecting Who: Artificial Beings*, clockwork mechanisms became very sophisticated during the Enlightenment. It was during this period that France became the center for the construction of ingenious mechanical toys and automata. Jacques de Vaucanson (1708–1782) was said to be the most skilled engineer of all, and was believed to have a secret ambition to build an artificial man. Such ideas led French philosopher Rene Descartes to conclude that man and other living creatures were themselves forms of automaton occupied by a spiritual mind. So that's the kind of pedigree that "The Girl in the Fireplace" leans upon.

"A Christmas Carol" (2010): Eleventh Doctor
The original *Christmas Carol*, written by Charles Dickens way back in 1843, was a timey-wimey tale about a grumpy miser named Scrooge whose visions of past, present, and future changed his mind about being so grumpy. Likewise, in this tale, the Doctor travels back in time to change the mind of the grumpy boss of Sardicktown, but the Doctor finds that rewriting people is far trickier than rewriting time. A crafty timey-wimey retake on a classic old tale.

This story's science innovation:
Causative psychiatry. Kazran is unlike other villains found in *Doctor Who*. He's not *all* bad, but more of a "damaged" character.

The Doctor plays psychiatrist and tries to change Kazran's past to "defrost" his soul.

"The Girl Who Waited" (2011): Eleventh Doctor

A strange and timey-wimey "Doctor-light" tale in which Amy gets stuck in an alternate timeline and neither the Doctor nor Rory can save her. By the time they come back to save her it's too late. Years have passed and she's not the same person she was, she's damaged. The Doctor and Rory have to make a choice, they can only save one Amy, as only one can survive in the time-stream.

This story's science innovation:
Speculative time-travel sci-fi based on the Many-Worlds Interpretation (MWI) of quantum mechanics, which holds that there are many worlds that exist in parallel at the same space and time.

"The Name of the Doctor" (2013): Eleventh Doctor

The Doctor travels to Trenzalore, the place where he is buried. Discovering an ailing TARDIS (his tomb), the Doctor sees the Great Intelligence launch himself into his time-stream—the only thing left behind when the Doctor dies. Since the Doctor is being erased from history, Clara must jump into the time-stream after the Intelligence and fix things. Saving the Doctor's life, she sprinkles herself across his time-stream, which is why the Doctor meets Clara several times. Ingenious timey-wimey stuff, and based on the idea that if there's one place the Doctor could never go in space and time, it's his own grave.

This story's science innovation:
A simple kind of MWI of quantum mechanics, though reliant on a single time-stream. This transgresses the "self-consistency principle" for time travel, of course. The Doctor usually does.

"Time Heist" (2014): Twelfth Doctor

The Doctor turns bank robber when he's given a task he cannot refuse—to rob the most dangerous bank in the cosmos. After a mysterious phone call, the Doctor and Clara find themselves in the maximum security bank, memories wiped clean and with two strangers to accompany them. The rest of the plot hinges on the timey-wimey logic of the Moffat era. The Doctor sums it up when explaining how the Architect managed to get the suitcases into the bank for him to open: "By breaking into the bank in advance of breaking into the bank!"

This story's science innovation:
A cosmic time-travel take on the classic heist trope.

"Inferno" (1970): Third Doctor

When the TARDIS console malfunctions, the Third Doctor is accidentally thrust into an alternate reality, where events are slightly different from his own. Hard-headed scientists intend to penetrate the Earth's crust using a special drill, but soon discover that their actions will lead to the devastation of the Earth. Narrowly escaping being burnt to a cinder, the Doctor manages to escape the parallel Universe just in time to stop the drilling in this Universe. Thank heavens for parallel dimensions.

This story's science innovation:
Another kind of MWI of quantum mechanics, focusing on parallel worlds, rather than alternate timelines.

"Day of the Daleks" (1972): Third Doctor

A band of guerillas from the twenty-second century travel back in time to save world leaders from being blown up at an important meeting to avoid World War III. In the future, the Daleks had managed to enslave Earth's war-exhausted humans. The guerillas arrive in 1972 to blow up the man they think is responsible for the initial explosion. However, it turns out that the guerilla's bomb was the actual cause of the explosion that killed the world leaders. Meaning, they were actually the cause of the war they're trying to stop. Like Dalek *Terminator* meets *Planet of the Apes*.

This story's science innovation:
A great example of a time-travel paradox.

In What Ways Does *Doctor Who* View the End of Our World?

In the Ninth Doctor story "The Dalek Invasion of Earth" (2005), the story is set on the Earth in the twenty-second century, where the Daleks occupy the planet following a meteorite strike and a deadly plague. Whovians will be aware of the fact that this is not unusual. It seems that, in the long history of Doctor Who, aliens are forever trying to get the better of our planet.

The End of the World

According to *Doctor Who*, the end of the world is nigh. Just how close we are, and by what means our planet will meet its destruction, no one really knows. But you can bet that popular and scientific concern about the coming apocalypse has often been driven by science fiction like *Doctor Who*. In the past, religious depictions of apocalypse appeared in many faiths. In modern times it's science fiction like *Doctor Who* that has delivered an ongoing exploration of our demise. Just how this pale blue dot in space might be extinguished, or at least the technological apes living on it, is a question with very varied answers. Here are some ways in which *Doctor Who* has pictured the end times.

Plague Doctor

"The Ark" is a First Doctor story set around ten million years in the future. The Doctor and his companions arrive on a generation ship that they name "The Ark." A plague has spread across the human and Monoid races on the ship, so the Doctor looks for a cure. The end of the story is set seven hundred years later, and sees the Doctor trying to stop the Monoids from wiping out the last of humanity with a bomb! Elsewhere the history of *Doctor Who* is littered with plagues of Daleks and Cybermen pushing the Earth to the brink of extinction. "The Ark" reminds us of the real horror of a plague threatening to annihilate the human race during the Spanish Flu pandemic between 1918 and 1919. Around 5 percent of the world's population died, around one hundred million souls, making it one of the deadliest natural disasters in human history. *Doctor Who* needs little encouragement. The science fiction portrayal of future pandemic apocalypses has influenced the World Health Organization to plan and coordinate responses to future outbreaks across the world.

War Doctor

War is generally considered a bad thing. Mark Twain claimed "God created war so that Americans would learn geography," and Albert Einstein said, "I know not with what weapons World War III will be fought, but World War IV will be fought with sticks and stones." The list of war's death toll throughout history is truly jaw-dropping. To name but a few: The Three Kingdoms War in China between 220 and 280 AD claimed around thirty-seven million deaths; the Spanish Conquest of the Aztec Empire claimed about twenty-four million deaths; the Taiping Rebellion in China between 1850 and 1864 claimed another forty-five million deaths; while World War I claimed a "mere" twenty-four million. But the largest and deadliest war in human history to date was, of course, World War II, with an estimated seventy million deaths (twenty-seven million of which were Russian deaths). With so much war on Earth, it's little wonder that *Doctor Who* should also portray war in heaven, some of which was brought back down to Earth. The most impressive of these portrayals is the time wars. The Whoniverse boasted two types of time war. The first type were those conflicts where enemies warred across different times in history. The second type were those in which time itself was a weapon. In these conflicts, time-active antagonists could use warring devices such as preemptive strikes, time loops, temporal paradoxes, and even the reversal of historical events themselves. Much has been written about the falsification of Earth's history in fact. In *Doctor Who* fiction, it was near impossible to study details of the time wars, as war atrocities were often erased before they were set in stone.

Earth Crisis Doctor

There are plenty of stories about the demise of our world as we know it, natural as well as technological. In the Tenth Doctor story "The Stolen Earth," our planet is teleported out of its spatial location in the solar system. The Doctor contacts the Shadow Proclamation, a universal police force, to find Earth and twenty-six other missing planets. Down on Earth, the familiar sky is gone. The Sun is gone. The constellations have been replaced with strange new ones. And yes, twenty-six new planets have appeared in the sky. The Doctor discovers the twenty-seven worlds in a nebula region of space known as the Medusa Cascade, an inter-universal rift. The idea of moving the Earth also features in the 2019 Chinese sci-fi movie *Wandering Earth*, which follows a group of astronauts and space workers guiding the Earth away from an expanding Sun, while also trying to avoid a collision with Jupiter. (By the way, *Wandering Earth* is China's third highest-grossing film of all time, the third highest-grossing non-English film to date, one of the top twenty highest-grossing sci-fi films so far, and has been described as China's first full-scale interstellar spectacular.) Whereas the prospect of shifting us into deep space will remain the preserve of sci-fi, moving the Earth farther away from the sun has actually been considered in science "fact." A number of planetary scientists, including some at Cornell University, have done some serious research into moving the Earth. As the solar luminosity increases over the next billion years, astrophysicists have been considering various mechanisms to increase the size of the Earth's orbit.

Alien Invasion Doctor

Watching *Doctor Who*, you could be forgiven for believing that aliens are about to land some time very soon. So deep is the *Doctor Who* conviction that beyond the dark, the cosmos is so teeming with life that they will come across space at any moment, perhaps tomorrow. Later, when the thrill dies down and you contemplate the infinity of time, you wonder if the Daleks came long ago, sticking in the swamp muck of the steaming coal forests, their bright robotic shells being clambered over by hissing reptiles, and their delicate instruments bitten off by irate raptors. Daleks versus dinosaurs. Of course, aliens invade Earth an awful lot in *Doctor Who*. And it's not just sadists like Daleks and Cybermen who want to dominate or destroy our planet. Plenty of other alien races have been hostile to humans, including Ice Warriors, Sontarans, the Autons, and, naturally, the Slitheen. As the Whoniverse probably has trillions of planets you do wonder why aliens bother coming all this way to planet Earth? The truth of the matter is, the Doctor is a dangerous ally for humans to have. Without him/her, how would they know about meek little Earth in the first place?

Cyber Doctor

Not content with destroying planet Earth in this Whoniverse, *Doctor Who* even has plans for Earth in parallel Whoniverses too. Consider the Tenth Doctor story "Rise of the Cybermen"/"The Age of Steel." In the story, in which the Cybermen make their first return since the show's 2005 revival, businessman John Lumic has overthrown the British government and taken over London in a parallel Whoniverse. With the Doctor's help, a human resistance movement seeks to stop Lumic's plan to "upgrade" all of humanity into Cybermen by placing their brains inside metal exoskeletons. Key to the resistance is the destruction of a transmitter Lumic has installed in London to control its population.

Sure, Time Passes;
But How Does It Do So on The TARDIS?

"In comparison with the distances light travels, all distances in the dimensions of space, even those involving airplane travel, are so very small that we essentially move only along the time axis, and we age continually. Only if we are able to move away from our frame of reference very quickly, like the traveling twin . . . would the elapsed time shrink to near zero, as it approached the speed of light. Light itself . . . covers its entire distance through space-time only in the three dimensions of space . . . nothing remains for the additional dimension . . . the dimension of time . . . because light particles do not move in time, but with time, it can be said that they do not age. For them "now" means the same thing as "forever." They always "live" in the moment. Since for all practical purposes we do not move in the dimensions of space, but are at rest in space, we move only along the time axis. This is precisely the reason we feel the passage of time. Time virtually attaches to us."

—Jürgen Neffe, *Einstein: A Biography* (1956)

TARDIS

Why is it called TARDIS? Sure, you probably know that the name TARDIS was invented by Susan (birth name Arkytior), who was an original companion and granddaughter of the First Doctor. And you no doubt know that TARDIS stands for **T**ime **A**nd **R**elative **D**imension **I**n **S**pace. Let's put to one side for a moment all this stuff about relative space and time dimensions. After all, TARDISes are meant to move through time and space by disappearing here and reappearing there, which they do using a component called the "dematerialization circuit."

What's of more interest to us in this chapter is how time ticks on the TARDIS. You may recall earlier on page 76 when talking about the Blinovitch Limitation Effect, we mentioned that one of Albert Einstein's great realizations was that time is not absolute. We also spoke of the famous adage of popular relativity theory, *moving clocks run slow*.

How does this help us answer the question as to how time ticks on the TARDIS? It's associated with Einstein's teaching that any measurement of time uses the idea of simultaneous events. All our judgments in which time plays a part are always reckonings of simultaneous events. For example, if the Doctor says, "The TARDIS arrived here at ten o'clock," he or she means, approximately speaking, "The small hand on my analogue watch pointing at ten and the arrival of the TARDIS were simultaneous events." In short, Einstein argued that simultaneous events in one frame of reference would not necessarily be simultaneous when viewed from another frame. Einstein called this "the relativity of simultaneity."

Relativity of Simultaneity

Let's do a thought experiment. Einstein was famous for them, so it seems fitting. Warning: this will be a weird thought experiment. After all, this *is* Doctor Who. In our experiment, the TARDIS is seated *exactly* in the middle of a passenger railway car with glass sides. (I told you it would be weird.) The TARDIS and the railway car are the moving frame of reference and we are seated on a railway station platform, which is the stationary frame of reference. The train that carries the TARDIS and the railway car is passing through the station on which we're seated.

Now imagine the TARDIS sends out a pulse of light in the forward direction of the glass-sided car, and at the same time sends out a pulse of light in the backward direction. Let's also imagine that the doors at each end of the railway car open automatically when the light pulses arrive. To the Doctor in the TARDIS, both doors of the railway car will open at the same time. That is, simultaneously. This makes sense. The TARDIS is seated equidistant from each door. And as the time taken to reach each door is given by distance divided by speed, where the distances are equal and the speed is the speed of light, the time of each door opening will be the same.

But to us, seated on that railway station platform, the back door will open before the front door. That is, *not* simultaneously. Don't believe me? Think about it. For us, stationary as we are, we see the back door *move forward* to meet the light pulse from the TARDIS. And we see the front door *move away* from the light pulse. This also makes sense. As the front door *moves away* from the light pulse, the pulse now has to go a *greater* distance on its journey than the backward light pulse does. That's because the back door is *moving toward* its light pulse, making a *shorter* distance for its journey. And, as the distances are not the same, the journey times to front and back doors will be different. In short, the doors will *not* open simultaneously.

But which observer is correct, I hear you scream? Do the doors open simultaneously, or not? At the risk of infuriating the reader even further, I am duty bound to say that *both* observers are correct. This "simple" example proves Einstein's point. Simultaneous events in one frame of reference (the Doctor in the TARDIS on the train) would not necessarily be simultaneous when viewed from another frame (we observers seated on the platform). You may need some time, and perhaps a stiff drink, to get used to this.

Tick TARDIS Tock

Hopefully, you are beginning to see that space and time measurements are relative. They depend on the motion of the observer. But, once more, how does this help us answer the question as to how time ticks on the TARDIS? It's in the understanding that the "clocks" being used, both on the train and TARDIS and on the railway platform, are both behaving normally. If the moving TARDIS clock changed when the Doctor was moving, then he or she could tell they were moving by noticing the clock change and that's simply too weird.

No, the answer is all in the fact that the stationary observer, us, sees something completely different when they look at the moving clock. As the speed of light must be the same for all observers, we stationary observers must hear more time elapse between ticks of the moving TARDIS clock than on our stationary clock. In other words, moving clocks tick slower than stationary ones, stretching out the interval between ticks and effectively meaning that each moving second is longer than a stationary second. And that not only explains how time ticks on the TARDIS, but also helps explain time problems such as the Twin Paradox.

Could the Doctor Wormhole through Our Universe?

In the Tenth Doctor story "Planet of the Dead" (2009), the story features a distant planet with mysterious alien sand, three Suns, and a wormhole.

"How do you decide whether tomorrow's technology includes time travel? Where do you look for evidence that our descendants have discovered the means of temporal voyaging? If time travel is a one-way process forward, there is no way we can know. If, as the new physics suggests, it is possible to move back in time, then the evidence we are searching for will present itself as anachronisms. Human beings are careless. They drop things they shouldn't, like the metal tubes found in Saint-Jean de Livet in France. They are also vulnerable. Whatever safeguards are in place, sooner or later someone will be trapped in a time period other than their own and die there. If the time period is historical, their death will leave no anachronistic trace, but if we examine the depths of prehistory, it becomes possible to trace the series of temporal disasters, which left a trail of corpses where they decidedly should not be."

—J. H. Brennan, interview, *Time Travel: A New Perspective* (1997)

Time Travelers

Picture this: the Doctor is standing at the Crucifixion. Captivated and awestruck, he or she can't help but study the scene. One of the most famous in all of history, at least on this world. Here is one of the benefits of time travel. Experiencing, firsthand, history in the unfolding, even if it is parochial to this planet. Just a few points to remember. He or she must do nothing to alter history. (Note to self: no messing with the Blinovitch Limitation Effect *this* time!) And when the crowd is asked who should be saved, she should join in with the call, "Give us Barabbas!"

Suddenly, the Doctor realizes something weird about the crowd. Not a single soul from 33 AD is here at the scene. The mob condemning Jesus to the cross is made up of time travelers from the future. And *some* of them from Mars!

But wait. The scene isn't *passively* littered with folk from the future. They've *actively* changed the outcome of history itself, by being present at the Crucifixion. The time travelers think they know the way history is meant to go. Rather than Jesus being set free, the crowd is meant to choose Barabbas, the bandit. But the decision only goes that way because travelers from the future are witness to the scene. Would Jesus have been set free instead, if they hadn't interfered? Bang goes the Blinovitch Limitation Effect again.

The above story is one of my favorite ways of introducing the idea of time travel and its contradictions. The tale can be told a little differently each time, depending on context. But credit must be given to the wonderful story upon which this introduction is based. This nifty time-travel scenario (not including the *Doctor Who* twist, of course) was dreamt up by Garry Kilworth in his 1975 science fiction story *Let's Go to Golgotha*. It's typical of the kind of conundrum that the *Doctor Who* writers have to reckon with when they tell a tale about tampering with time.

A Brief History of Wormholes

As you may know by now, the most scientifically credible way to achieve time travel is through the use of wormholes. Possibly. What do wormholes look like? You remember the title sequence at the beginning of many programs of *Doctor Who*? You get the bit where they tell you the name of the episode, along with the cast and production members who worked on that program. Then, as the unforgettable *Doctor Who* theme song spooks out of your speakers, you see the TARDIS falling through what looks like a sinkhole in space. Or like a giant plughole is sucking the TARDIS into oblivion. That sinky, pluggy thing is a wormhole, as the Tenth Doctor might say.

They're called wormholes because they're a little bit like the tunnels that worms make in apples when they eat their way through. The journey through the apple tunnel is shorter and quicker than going over the apple's surface from one side to the other. That's how wormholes work in space too. Einstein figured there would be regions of space where there are shortcuts, linking vastly separated regions with a much shorter tunnel.

Famous American science-fiction writer John Campbell was the man who invented such tunnels or "space warps." In his 1931 story *Islands of Space*, Campbell used the idea as a shortcut from one region of space to another. And in his 1934 story *The Mightiest Machine*, he called this same shortcut "hyperspace," another now-familiar phrase. A year later, world famous Nobel Prize–winning scientist Albert Einstein (him again) along with colleague Nathan Rosen, came up with the fact behind the sci-fi invention of time travel. Einstein and Rosen worked out the scientific theory that explained the notion of these "bridges" through space. It was much later that we all started calling these bridges "wormholes."

So *that's* why you often see the swirly cosmic tunnels of wormholes in many movies. It's meant to show when a spacecraft is on a journey through space and time. Some character in the movie might even do you the favor of saying that a wormhole has at least two mouths, connected to a single throat, or something like that. They may even grace you with the fact that scientists really do believe they exist. At least in theory. And, as that theory is Einstein's, people take it seriously. Especially sci-fi writers. Stuff may "travel" from one mouth to the other by passing through the wormhole. We haven't found one yet. But the cosmos is immense, and we haven't really been looking for very long. Please also bear in mind that American cosmologist Lawrence Krauss, author of the 1995 book *The Physics of Star Trek*, suggested on NBC's *What Einstein and Bill Gates Teach Us About Time Travel* in May 2017, "Most physicists now working would bet against the possibility of time travel, not merely because of the practical difficulties of generating the necessary conditions to allow it, but also because of the implications of time travel if it becomes possible."

Wormhole-ing The Doctor

So how could wormholes help the Doctor travel in *this* Universe? Imagine we create a wormhole. As we have said, a wormhole is a region of space-time that's warped. That is, a "shortcut" in space and time through which to travel. The trouble is, according to science fact, the Doctor would not be able to travel back in time to a date before the wormhole was created. For example, if the Doctor managed to create a wormhole on April 1, 1999, he or she wouldn't be able to go back in time before 1999. So, in order to travel back further, some splendid Time Lord in the distant past would have had to conjure up a wormhole to get the whole thing going.

Here are two wormhole recipes for the Doctor. Recipe one: the Doctor takes a dash of exotic matter. He makes sure this matter is made up of particles that have antigravity properties. He pops them into the throat of a wormhole. (Note: he avoids black holes, which are one-way journeys to oblivion. The Doctor's wormhole should have two mouths, an exit and an entrance. His challenge will be to keep the wormhole's throat open by using a force opposed to gravity, kind of antigravity, if you like.) Geronimo! The wormhole throat stops imploding, and the Doctor has made a wormhole. The stuff of sci-fi has become fact.

Recipe two: the Doctor manages to make a wormhole that has two mouths side by side. He plonks one of the mouths into the TARDIS. (It should be easy enough to take a wormhole mouth into

the TARDIS, as wormholes are areas of extremely warped space with huge gravity fields, which would be attracted to the TARDIS if he could convince the wormhole that the TARDIS was really massive.) Then, when the TARDIS goes worming away at great speeds, the time-frame of the TARDIS mouth has slowed down a lot, compared to the mouth of the wormhole he's left behind (that's because "moving clocks run slow," according to Einstein, as we know!). Now, if the TARDIS brings the traveling mouth of the wormhole back to the other mouth, let's say ten years later, then the TARDIS mouth would have effectively jumped ten years into the future!

What's the History of *Doctor Who* in Ten Objects?

"Ah, the Sonic Screwdriver, that most divisive object in the New Who arsenal. First introduced at the end of the sixties, it was written out of the show in 1982 at the instruction of producer John Nathan-Turner. Along with several other members of the production crew, Nathan-Turner felt that giving the Doctor a device that let him unlock doors, weld metal, detonate explosives, and fix machinery limited the writers' dramatic options—so the old faithful got zapped by a Terileptil in 'The Visitation' and the series soldiered on without it."

—Tim Martin et al., *Not everybody loves the Sonic Screwdriver*, in *The Daily Telegraph* (2015)

A History of Things

In 2010, the British Museum and the BBC decided to grapple with Earth history. After four years of planning, and presented over a period of six months on BBC Radio 4, they detailed one hundred objects of past art and technology, all of which are in the British Museum, that tell a history of the world. The one hundred objects ranged from a Tanzanian million-year-old hand axe, and the statue of a Minoan bull-leaper from the ancient culture on Crete, to the ship's chronometer from Darwin's HMS *Beagle*. The Museum suggested that this landmark project told "a" history, not "the" history.

So, what better way to give a sense of time about the influence of *Doctor Who* on modern culture than to set up our own thought experiment? For want of space, we shall reduce our list from one hundred objects down to a magical ten. But we will retain the same intention to briefly tell the chronological tale of each artifact from "a" chosen history of the Doctor and his travels. Each entry is accompanied by a "science gravitas" footnote, where we try to weigh up the deep and philosophical resonances of the artifact concerned.

Pandorica

The Pandorica was built by the Alliance to stop the Doctor from inadvertently destroying all of creation and existence itself. According to legend it was the prison of a warrior or goblin who dropped out of the sky and tore the world apart, until a good wizard tricked it and locked it up.

Science gravitas: the Pandorica was a prison hidden under Stonehenge, a world heritage site constructed over five thousand years ago and, like the Doctor, a British cultural icon. How's that for double-whammy gravitas? Downside: despite being described time and again by the Eleventh Doctor as the "perfect prison," the Pandorica was easily opened using, yes, you guessed it, the Sonic Screwdriver.

The Moment

Also known as the Galaxy Eater, the Moment was the most dangerous and powerful weapon in all of creation. The Moment was able to breach time locks and create tears in the fabric of creation known as time fissures. The Doctor had intended to use it to end The Last Great Time War. Built by the ancients of Gallifrey, its sophisticated operating system was so advanced it became sentient and took the form of Rose Tyler.

Science gravitas: Joining the long list of sci-fi superweapons, the Moment got its name as it was capable of destroying whole galaxies within a single moment. Almost like an anti–big bang. Kind of.

Untempered Schism

The Untempered Schism was an opening in space-time, a gap in the fabric of reality through which can be seen the whole of the

Time Vortex. Like all Time Lords, the young Master was taken for his initiation ceremony at the age of eight, during which he gazed through the Untempered Schism, and went bonkers. From then on he heard constant drumming, which worsened with time. As a sufferer of tinnitus, I'll bet that drumming can get on your nerves after a while. It's said that exposure to the Schism over billions of years is what gave the Time Lords their ability to regenerate, and was a major influence in their evolution.

Science gravitas: One can think of stargazing as a kind of Untempered Schism. Stargazing is certainly a window on the space-time fabric of reality, and since the further out you look the further back in time you see, it's also a kind of Time Vortex, though admittedly not very vortexy.

Time Television

The Time Television, also known as the Space-Time Visualizer, was a device that allowed the observer to watch any event in history. With an origin in the misty history of the Whoniverse, these televisions were used to watch Queen Elizabeth I, William Shakespeare (whom my late father used to call Billy Waggledagger), and The Beatles.

Science gravitas: The Time Television could tune in to events all over the Whoniverse by converting neutrons of light energy (utter scientific nonsense) into electrical signals. Sadly, the televisions only seem to come in black and white, in perhaps one of the most profound technological mismatches in sci-fi history: able to tune into the Universe, but only in monochrome.

The Sash of Rassillon

The Sash is one of four important items inherited and owned by each new president of the Time Lords. It was initially fashioned by the Solar Engineer Time Lord, Omega, to allow him to not be destroyed by the Eye of Harmony, which he also created.

Science gravitas: Can protect the wearer from being spaghettified by the awesome forces surrounding a black hole. Science: *nothing* can protect you from being spaghettified by the awesome forces surrounding a black hole.

The Key to Time

This perfect cube (what cube isn't perfect?) was the purpose for the entire sixteenth season of *Doctor Who*. It contains the elemental force of the Whoniverse and is said to maintain the equilibrium of time itself. It consists of six segments, scattered and hidden throughout time and space. The Doctor's mission was to locate them. When they are assembled into the cube they provide God-like powers, which are too dangerous for any being to possess. It's so powerful that it can be used to stop the entire Whoniverse. *Science gravitas*: Joins the long list of sci-fi superweapons (see the Moment on page 165).

TARDIS Key

Is the TARDIS key a significant artifact? It's the key to the exterior door of the Doctor's space-time machine, so that's pretty significant. It can be found in a variety of shapes and designs, but the key goes far beyond that. The key is Gallifreyan technology that opens other TARDIS doors, glows when the TARDIS is about to materialize, acts as a kind of remote that brings the TARDIS back, and can only be destroyed by being dropped into lava.
Science gravitas: The Whoniverse equivalent of a universal remote for domestic appliances, or a car key fob.

Handles

"Handles" was the name given by the Eleventh Doctor to a cyberman head that he'd gotten from the Maldovarium Market. Once all the scraps and electrics had been scooped out, the Doctor found himself with a docile robot head, which would obey his commands.

After many adventures, Handles began to corrode and lost all function, reducing the Doctor to tears. Handles is the longest-serving companion the Doctor has had, as he was on Trenzalore with the Doctor for the first three hundred years.

Science gravitas: Handles has similar qualities to Amazon's Alexa (such as reducing you to tears), only less annoying.

Mona Lisa Multiples

Anything associated with the *Mona Lisa*, Earth's most famous work of art, is bound to be weighty. *Doctor Who*'s version of the painting is that it was commissioned in multiples (i.e. there were six forgeries made). Except you can't really call them forgeries as Leonardo da Vinci painted them all. The six copies were stored away in a secret Parisian cellar until 1979.

Science gravitas: The Fourth Doctor ensured that future X-ray machines would be able to identify the copies as they have the writing "THIS IS A FAKE" written somewhere on the work. ·

The Matrix

Similar to the popular movie franchise but decades earlier, the Matrix is a massive computer system on Gallifrey that can provide a simulated cyberspace. It's kind of like a Time Lord virtual reality system where virtual worlds can be created. Not only can it be accessed by wearing tech, such as the Crown of Rassilon, but it can also be physically entered via an entrance known as the Seventh Door.

Science gravitas: It holds the combined knowledge of all past and present Time Lords, including the imprints of their personalities. This idea appears to predate many sci-fi stories about the transcendence of consciousness via computer uploading.

Is There a Problem with the Doctor Traveling Faster Than Light?

In the Third Doctor story "The Claws of Axos" (1971), the Doctor uses a stream of Axonite particles and accelerates them through time to achieve light speed.

"The telescope, in enabling us to look far out into space, also allows us to look back in time. Light travels at about 186,000 miles per second. When we look up into the daylight sky, we are not seeing the sun as it currently is but as it was about eight minutes ago, since it takes that long for the light radiating from this familiar star to travel ninety-three million miles to Earth. Similarly, when the Giant Magellan Telescope (GMT) receives light waves from the depths of the Universe, those waves will have originated from points as far as seventy-six sextillion (76,000,000,000,000,000,000,000) miles away. It will have taken those waves some thirteen billion years to arrive on earth, meaning they left their source about a million years after the big bang, and roughly nine to ten years before Earth even formed."

—Richard Kurin, *The Smithsonian's History of America in 101 Objects* (2016)

Taking Off in the TARDIS

Imagine you're standing over the control console of the TARDIS. With so much power at your fingertips, you wonder if it might be possible to take the TARDIS beyond the speed of light. What are your chances? And how does the Doctor do it?

To answer these questions, we first have to think about what happens when you try to get an object like the TARDIS to go faster than light. To get the TARDIS or any other object moving in the first place we have to apply a force. Now by force we don't mean "strength" or "an armed body of men." We mean a force in physics, like a push or a pull or a kick. (Not that kicking the TARDIS will send it off at faster-than-light speeds.)

In science, the idea of a force is the idea of interaction. So, to get the TARDIS moving, we need to give the craft a series of hits, or maybe a steady stream of pushes, for example, by some kind of engine. Of course, on Earth there are a number of challenges to applying a big steady force. Air resistance, for example, which admittedly won't trouble us in space! Although running out of fuel and mechanical breakdown are concerns on Earth and in space alike.

Talking Inertia

But there are bigger challenges than those. That's because there are no instantaneous interactions in nature. Everything takes time. And this has a bearing on what happens when an object starts to approach light speed.

Imagine we apply a steady force to the TARDIS. Don't worry so much about *how* we're doing this, just that we *are* doing it. When the TARDIS then picks up speed we say it accelerates. British physicist Isaac Newton was one of the first to realize the link between force and acceleration. In fact, Newton put it something like this: the acceleration of a body is in proportion to the force applied to it, and inversely proportional to the body's mass, which is also sometimes called the inertia of the object. Now, these last points are of interest. They mean that the bigger the force, the faster an object picks up speed. But also that the bigger the body mass or inertia the harder it is to get moving quickly (this makes sense as it's easier to get a minicar going than a monster truck).

It was Albert Einstein, of course, who showed that mass is not constant. A body like the TARDIS has a rest mass, the mass when it is stationary. But, as the speed of the TARDIS increases, its inertial mass also increases. For low speeds, those much less than light speed, this increase in mass is barely perceptible. But as the TARDIS gets close to the speed of light, its mass starts to increase swiftly toward infinity. In theory, the TARDIS mass would become infinite if it accelerated all the way to light speed. But, because the acceleration of a body, like the TARDIS, in reaction to an applied force is inversely proportional to its inertial mass, as light speed is approached the force needed to reach light speed also becomes infinite. And that's why it's not possible to accelerate the TARDIS to the speed of light. In short, and in the spirit of Einstein, if you gave the TARDIS more and more energy, instead of going faster and faster, it just gets heavier and heavier. No matter how much thrust you gave it, the TARDIS would still be going less than light speed.

Final Footnote

All this is known to the writers of *Doctor Who*, of course. Adric, companion to the Fourth and Fifth Doctors, reminded us that the speed of light is the subject of Einstein's Theories of Relativity and involves $E = mc^2$, the most famous equation of the twentieth century. Even though Einstein said it wasn't possible to travel faster than light, the Third Doctor seemed to think it was and even did experiments to prove it, as in the story "The Claws of Axos."

In the 1996 *Doctor Who* movie, the Eighth Doctor says that Gallifrey is 250 million light-years away. That means it should take 250 million years to travel from Earth to Gallifrey at the speed of light. And yet, in the TARDIS, the Doctor's journey is made in a matter of minutes. How come? Apparently, the trick is to take a different track than the light beam. And the track the Doctor takes in the TARDIS is the Time Vortex. Yeah, right.

Is the Doctor Right about Alternate Timelines and Parallel Worlds?

In the Tenth Doctor story "Rise of the Cybermen" (2006), the Doctor is in the parallel Universe's version of London. In the story, businessman John Lumic wants to upgrade all humans into Cybermen by placing their brains inside metal exoskeletons.

"Jonbar Point: term used for a crucial forking-place in time, whose manipulation can radically affect the future that follows. The name derives from Jack Williamson's *The Legion of Time* (1938), which deals with the potential future empires of Jonbar (good) and Gyronchi (bad). The former is named for the character John Barr: the fiercely contested jonbar point is the moment when as a small boy Barr picks up either a magnet, inspiring him to a life of science which ultimately brings Jonbar into existence, or a pebble, leading Barr to obscurity and the world to Gyronchi."

—John Clute and Peter Nicholls,
The Encyclopedia of Science Fiction (1979)

Other Worlds, Different Dimensions

Remember the *Doctor Who* episode "Journey's End" when a human version of the Tenth Doctor went to live with the real Rose Tyler, but in a parallel world with a different timeline? Well, Whovians with a nerdy eye for research might wonder if it's at all possible for such parallel worlds to exist. In a 1998 BBC book, *The Face of the Enemy*, Sergeant John Benton of UNIT described a parallel Universe as, "some sort of mirror Universe, like in that *Star Trek* episode where Spock had a beard." The Master, amused by this description, went on to quip that a parallel Earth occupied "the same space-time coordinates as this Earth, but in a different dimension. Sideways in time, if you like, rather than forward or back." Can all this possibly be true?

Science fiction history is full of such alternate worlds and parallel dimensions. It's all part of the sci-fi obsession with alternate history. You know the kind of thing: What if history had happened differently? What if grass had died? What if Neanderthals had not become extinct? What if a series of alien invasions had made the world well aware of life on other planets before the Renaissance? What if Christopher Columbus never sailed west? What if there was no oil in the Middle East? What if the South had won the US Civil War?

English author L. P. Hartley once said, "The past is a foreign country; they do things differently there." But in the sci-fi sub-genre of alternate history, things in the past *did* happen differently, and the present becomes a foreign country. Kind of. Alternate history stories in *Doctor Who* contain "what if" scenarios, like the list above, which revolve around crucial points in the past that could have had a different outcome to that recorded in actual history.

A Brief History of Alternate Histories

One early use of the alternate history idea crops up in an unexpected place. It's in Edward Gibbon's classic work *The Decline and Fall of the Roman Empire* (1776–1789). In his work, Gibbon speculates what might have happened had Muslims won the battle of Poitiers (it was a close call) in France in 733 AD: "The Arabian fleet might have sailed without a naval combat into the mouth of the Thames. Perhaps the interpretation of the Koran would now be taught in the schools of Oxford, and her pulpits might demonstrate to a circumcised people the sanctity and truth of the revelation of Mahomet."

The Book of Mormon (1830) could also be seen as alternate history. According to *The Book of Mormon*, Old World migrants came to the Americas in several waves, mainly Jews from the Levant, and inhabited the continent from about 2,000 BC to 400 AD. Allegedly, these mysterious migrants built sumptuous cities large enough to support hundreds of thousands of warriors, and Native Americans are mostly descended from them. *The Book of Mormon* is a variation on a commonly held belief in early nineteenth-century America—namely, that the Americas was settled by Old World immigrants whose established and advanced culture fell into decline. Conveniently, all evidence of this culture had mystifyingly vanished by the arrival of Europeans in the 1490s. Understandably, *The Book of Mormon* narrative is regarded as nonhistorical by a wide range of scholars. To quote just one single piece of real science on the matter, the Native American nation known as the Navajo have a gene marker inherited from the Chukchi, a tribe that still lives deep in Arctic Russia. Scholars believe the Chukchi were among those who migrated from Asia to America, many millennia earlier than 2,000 BC.

Doctor Who is not alone. For most of the twentieth century, modern cultural accounts of alternate history have been associated with sci-fi. Often, these histories are infused with time travel. Jumping from one history to its alternate, as in "Rise of the Cybermen," an awareness of the presence of one timeline by the people in another is a common theme of the genre. Indeed, cross-time alternate histories have become so closely related that it's almost impossible to separate them from the genre as a whole. Perhaps history's most influential early sci-fi alternate history is H. G. Wells's *Men Like Gods* (1923). The tale features travelers transported to an alternate Earth, as in *Doctor Who*, which is almost a utopia, and which has diverged from our own history several centuries before. As a result of Wells, timeline-hopping tales became incredibly popular in the twentieth century.

Steampunk Doctor

Another notable alternate history is *The Difference Engine*, by William Gibson and Bruce Sterling. The novel details a history in which Charles Babbage, the patron saint of the programmable computer, transforms the Victorian world with his inventions. London becomes a proto-cyber-city as the result of the successful production of Babbage's analytical engine, an early computer that didn't quite make the grade in our own timeline. Gibson and Sterling's counterfactual fiction imagines the mass production of card-driven computers that transform western society and fuel a turbo-charged industrial revolution. These difference engines lead to a changed world, and the information revolution of the late twentieth century is played out against a steampunk setting.

The history of *Doctor Who* is full of steampunk. Especially since Gibson and Sterling wrote *The Difference Engine*. Take the Eleventh Doctor story "The Snowmen," for example. The Doctor is in nineteenth-century London, dealing with snowmen who intend to freeze the whole world so they will never melt. The steampunk comes in the form of aliens in bustles, Tesla-like lightning strikes, and references to Sherlock Holmes. Then there's the Twelfth Doctor story "Deep Breath." This tale could even be described as steaming-coal-forest-punk, as a dinosaur appears in the Thames. Meanwhile, as the dinosaur tries to dine out on a nineteenth-century Big Ben, the story also has a steampunk robot operated restaurant. Finally, as steampunk is most commonly associated with the Victorian era, consider the Tenth Doctor story "The Next Doctor." It's the year 1851, and a string of mysterious murders point toward an invasion of the Cybermen. But the steampunk is ramped up further by the inclusion of hot-air balloons, child-powered steam engines, and the brass-face-plates of cyber-cats, in what add up to a near-perfect steampunk setting.

Gibson and Sterling's alternate history is similar to the Doctor's occasional world of the black tarmacadamed streets of Victorian

London. Such stories conjure the great city into an even grimier smog-stained version of the metropolis than our own history. Though their take on those Dickensian days seems strikingly similar to our own time. In *The Difference Engine*, a distance between the timelines is cleverly struck with their "clackers," rather than "hackers," and their mean London back alleys that invest their world with considerable swagger. In *Doctor Who* and *The Difference Engine*, these aren't dull Victorians, they're Victorians with attitude. But behind the sci-fi swagger, the writers are making a serious point. They are pushing the origin of steampunk back into the Victorian past. They're relocating the matrix back into those days of Dickens, when change thrummed along electric wires, and the ubiquity of the modern computer finds its antecedents in the pioneering science of the Victorian era. In the Ninth Doctor story "The Unquiet Dead," Dickens even makes an appearance, though the tale is more ghostly than steampunk. (The introduction of gas lighting in the Victorian era may have had something to do with this upsurge in ghost sightings and spiritualism. Carbon monoxide poisoning may have caused large numbers of people to hallucinate.)

Science Gets Down with the Doctor

Some scientists believe *Doctor Who* may have a point. As our Universe expanded rapidly after the big bang, it may have formed into a vast number of disconnected bubble Universes. An alien living on a planet in a galaxy one hundred billion light-years away is in a bubble so far away that it would be another eighty to ninety billion years before they would even be able to see our bubble through their most powerful telescopes. So, as these bubbles are so far apart, the worlds within them are, in effect, living in separate parallel Universes.

In fact, some scientists have gone so far as to work out that there are so many bubble Universes that there's likely to be an exact copy of you somewhere. These scientists believe there is an infinite number of other inhabited planets, including many that have the same face, body, name, and thoughts as you, and where your life choices are played out in an infinite number of ways. Perhaps on a parallel world *you* wrote this book. Or perhaps you're one of the key writers for a parallel world of *Doctor Who*. Or maybe you are of the opposite sex. But could you really jump from our own world to one of these parallel worlds, maybe through a wormhole?

In the words of Ben H. Winters in the 2016 alternate history *Underground Airlines*:

> Sometimes it's possible, just barely possible, to imagine a version of this world different from the existing one, a world in which there is true justice, heroic honesty, a clear perception possessed by each individual about how to treat all the others. Sometimes I swear I could see it, glittering in the pavement, glowing between the words in a stranger's sentence, a green, impossible vision—the world as it was meant to be, like a mist around the world as it is.

In What Way Do Humans Regenerate like Time Lords?

"Man's consciousness regarding generation and regeneration in Nature may have begun as early as the days of Paleolithic cave painting thirty to thirty-two thousand years ago, although this hypothesis is still controversial. Cave paintings from the Paleolithic era in France, Spain, and Australia, show human hands with missing digits beautifully painted on the cave walls. Various hypotheses suggest that the missing digits may be the result of frostbite, or ritual amputations, while critics maintain that the hands with missing digits are fraudulent representations placed in the caves by pranksters."

—Andrew Maniotis, *The Age of Regeneration* (2019)

Doctor Regeneration

What do these things have in common: you're wearing a bit thin; you're forced to change the way you look by court order; you've been poisoned by radiation; you fell from a great height; you absorbed time vortex energy; or you simply got to old age. Yes, that's right; they're all reasons why the Doctor regenerated.

The Doctor, like other Time Lords, can delay his death. Regeneration allows him to rejuvenate every cell in his body, and totally change the way he looks from time to time. It's an ingenious idea for a TV program, and it's probably the most important plot device that has kept *Doctor Who* going for more than fifty years. After all, how else would you be able to keep the same central character over all those decades? James Bond had to change actors. But *Doctor Who* had a narrative device ready and waiting for the change.

Human Regeneration

According to one prominent scientist, humans have their own programs of mini-regeneration. Whatever your age, your body is many years younger. In fact, even if you're middle-aged like me, most of your body may be just ten years old, or less. The simple facts of the matter are this: Most of our body's tissues are under constant renewal. This way of thinking about the age of human cells has been invented by Swedish Professor of Stem Cell Research Jonas Frisén. He believes the average age of all the cells in an adult's body may turn out to be as young as seven to ten years. So, in some ways, we *are* regenerating.

Professor Frisén's research suggests that cells from the muscles of the ribs, taken from people in their late thirties, have an average age of fifteen years. And his method is down to estimating how long cells actually last. The average age of the cells in the main body of the human gut is about sixteen years. The red blood cells last only 120 days. And an adult liver probably has a turnover time of between three hundred and five hundred days. In fact, 98 percent of atoms in the human body are replaced yearly, and the entire human skeleton is thought to be replaced every ten years or so in adults.

Of course, we humans don't have the same kind of variability available to the Time Lords. They can sometimes choose what they want to look like. It's happened at least twice on *Doctor Who*. The first occasion was the Second Doctor story "The War Games." (The Second Doctor was quite fussy about his next look and appearance.) Another occasion centered on the character Romana, another Time Lord from the planet Gallifrey, and a companion to the Fourth Doctor. Romana tried on a few bodies before settling on her eventual form. (These Time Lord facts make one wonder whether they have to stick to humanoid form, or could possibly transform into something completely alien?)

Back on Earth, regeneration is also a relatively common thing in the natural world. For example, when insect larvae undergo a dramatic change to morph into things like flies, moths, or butterflies. Indeed, caterpillars even have to digest themselves first to do so. Earth also has a few other tricks up its evolutionary sleeve. Take, for instance, the Axolotl, a.k.a. Mexican salamander. These amphibians have the ability to regenerate body parts, just like the Tenth Doctor re-growing his hand while regenerating. Pretty handy.

What Would Really Happen If You Lived as Long as the Face of Boe?

In the Tenth Doctor story "New Earth" (2006), the Doctor meets the mysterious Face of Boe. The story is set five billion years in the future on the planet New Earth, a planet humanity settled on following the destruction of Earth. It is suggested that the Face of Boe is thought to be millions, if not billions, of years old. But, when this was put to Boe, he merely replied with a leading question concerning the impossibility of such an age.

> "Personally, I would not care for immortality in the least. There is nothing better than oblivion, since in oblivion there is no wish unfulfilled. We had it before we were born yet did not complain. Shall we whine because we know it will return? It is Elysium enough for me, at any rate."
>
> —H. P. Lovecraft, *Selected Letters V* (1934–1937)

The Face of Boe

Could you live as long as the Face of Boe? If you recall the storyline in *Doctor Who*, the Face of Boe was one of the oldest creatures in the Whoniverse. By the end of his life, Boe was little more than a gigantic humanoid head. How did he last so long? Well, that very much depends on whether you believe the Face of Boe and Jack Harkness are one and the same person. If you recall, Jack was killed by a Dalek, but was then lucky enough to have been saved by Rose Tyler, who accidentally turned him into a being that ages super slowly. But how would mere mortals like us age as slowly as Lucky Jack?

Your body ages because it's fighting a constant battle between the forces of harming and healing. One of the harming forces is "free radicals," body chemicals that are very reactive with other chemicals. Sure, "free radicals" do good stuff in your body, like killing bacteria. But they can also do damage to the DNA in your healthy cells. As a result of this damage, your body has evolved a healing force that tries to limit the harm done by the "free radicals." In fact, every day in each cell of your body, the DNA is zapped more than ten thousand times by oxygen free radicals. Most are healed, but the few that aren't mount up and eventually cause the aging of your body.

One future way of dealing with this daily damage could be to use robots. Swarms of tiny little critter robots, called nanobots, would be injected into your bloodstream, a bit like the nanogenes that pop up in the Ninth Doctor stories "The Empty Child" and "The Doctor Dances." Once they're in your body, these nanobots, which are only a few thousandths of a millimeter in size, could act like mini-healing machines. They would not only do the job of making sure you get all the nutrients you need from your food, but they'd also do the job of zapping all those harmful "free radicals." That way, the nanobots may help stop the aging process.

Immortal Boe

In the episode "New Earth," it is mentioned that the Face of Boe may be millions of years old. And in "Gridlock," it's even suggested he's billions of years old. (Though when such statements of immortality are suggested to Boe himself, he simply replies wisely about the sheer impossibility of reaching such an age.) Immortality, of course, is a sci-fi and comic book classic. The idea of immortality is usually wrapped up with notions of the elixir of life, and the fountain of youth, both pipe dreams of science fiction. But what immortality usually means, as in the case of the Face of Boe, is extreme longevity, freedom from ageing, and relative indestructability.

Writers and thinkers alike have always had doubts about immortality. It's often treated as a false goal. Just think of the perpetual punishment handed out to the likes of Sisyphus, the king of Greek mythology, who was forced to forever roll an immense boulder up a hill. Or the Wandering Jew, who taunted Jesus on the way to the cross and was then cursed to walk the Earth until the Second Coming.

Writers have rarely been keen on the social impact of the kind of immortality bestowed upon the Face of Boe. Many have suggested it would lead to a kind of social sterility—a society going nowhere, and subject to little, if any, change. Even those writers who took a brighter view chose a rather elitist conclusion of a few privileged immortals living in a world of mortals. And when George Bernard Shaw showed enthusiasm for universal longevity in his 1921 work, *Back to Methuselah*, Karel Capek, the sci-fi writer who introduced the word "robot," responded to Shaw by saying immortality would be an unmitigated nightmare, even for a single person.

So, you may well ask, what's so bad about living "forever" like the Face of Boe? To a five-year-old boy, one year is 20 percent of his life. To his twenty-five-year-old mother, one year is only 4 percent of her life. The same 365 days feel very different for different people. With modern advances in medicine, it's quite possible the boy could live for one hundred years, or 36,500 days. But imagine living for 36,500 years. If our boy lives for 36,500 years, then a single year could feel to him like a day.

Let's go back to the Face of Boe. Would the boy's emotions hold true through the likely boredom of living for millions of years? Maybe this is why Loki seems so wise about the passing of time. Humans might become very sad and lonely, knowing they have and forever will outlive everyone they have ever loved. But does the Face of Boe care about that? Perhaps he is the perfect immortal.

A Whoniverse of Immortals

But what if everyone were immortal? Research into biotechnology following the cracking of the genetic code has meant science really may be able to keep the spark of life alive. The recipe for doing this is a varied one, running from treatments and medicines such as eugenics and genetic engineering, all the way through to artificially extending life with the use of synthetic organs (or by becoming a Cyberman). Another option for immortality involves uploading human consciousness into new bodies. In sci-fi stories that consider this immortal future, the humans did this uploading so many times that they begin to see themselves as gods—a slippery slope that hasn't caught out the Face of Boe.

Imagine an Earth replete with such immortals. Our planet is only so big. Where would we all live? This kind of question was tackled in the 1966 sci-fi novel *Make Room! Make Room!* Written by American novelist Harry Harrison, the book explores the consequences of a huge population growth on society. Set in a future where the global population is seven billion (uh oh), the world suffers from overcrowding, resource shortages, and a crumbling infrastructure. The book also serves as the basis for the 1973 movie *Soylent Green*, where the film went one better by introducing cannibalism as a way to feeding the growing number of people.

And what about the immortal darker side of dating? Has the Face of Boe any experience of this? Imagine our immortal Boe falls in love with a companion once every one hundred years. That means our serial monogamist immortal would have ten thousand girlfriends in one million years. That's some challenge. And it would totally change the definition of what a meaningful relationship means. For one thing, how many of those ten thousand companions' names will he be able to remember? No wonder the Face of Boe looks so old and troubled!

The Memory of Immortals

Memory would also be a problem for immortals. People can rarely remember in detail what they did last year, and especially when they were five years old. Think about it: What proportion of your past have you forgotten? Unless you have a truly exceptional memory, probably the vast majority of your past is lost forever. And if we have problems recalling what we did when we were five, how much would we remember if we lived for one thousand years, or if we lived to one million years old? Human brain capacity is limited. And that means we simply don't recall all of life's minutiae, as our minds merely replace useless memories, such as answers to the trivial security questions banks always expect you to remember, with far more vital data. Such was the life of Boe.

Another dastardly dynamic with Boe's immortality would be down to Darwin. It's this: people have not always looked the same. According to the Darwin-Wallace theory of evolution, as women find taller men more attractive, then taller men would be more likely to mate and have kids. And that means more tall genes in the gene pool so, next generation, more kids will have the genes to be taller. Repeated iterations of this sexual selection over one million years would mean that average human height will be lots taller than today's average height. (No wonder all that's left of Boe is the head!) And this all assumes that humans aren't completely wiped out by natural disasters, such as an asteroid impact, a giant solar flare, or a global epidemic.

By most accounts, human ancestors were short, hairy apes. Now, we don't look like apes anymore. Although we could be called "naked apes" compared to our closest living relatives, we actually have as many hairs per square inch on our bodies as there are on a chimpanzee. The human body is completely covered in hair from head to toe. But, as many of those hairs are so fine and fair, they're just not visible to the human eye. And we haven't always looked this way, of course. Our appearance has slowly evolved and gradually thinned our body hair over time. Now, imagine like Boe you are the only immortal living in a world of mortals. Everyone would continue evolving, generation after generation, so that you will end up looking very different from everyone around you. After all, imagine we pulled off a Jurassic Park trick on a Neanderthal. Most modern humans would fail to make friends with the fella, and simply call the local Museum of Natural History.

And one more drawback of being an immortal like Boe: polytrauma. Immortality is one thing. But invincibility is quite another. And being immortal doesn't necessarily mean being invincible. It merely means you can't die. Immortality doesn't provide you with a warranty for body health. Think about a normal human body and how many scars it might have. And then think about how many permanent scars an immortal will have after living for just one thousand years! Again, it's little wonder all that's left is the tired Face of Boe.

Look at mutilation in the United States. In one year alone, there are almost one-fifth of a million amputation-related hospital discharges in the US, mostly due to illness or accident. That percentage seems low when you compare it to a total population of 325 million souls and a life expectancy of around eighty years. But if you've been alive for Boe's million years, the chance of you keeping all your appendages is woefully slim. In fact, you're more likely to be a victim of major trauma, or polytrauma, and your injury severity score (ISS), a kind of medical score given to assess trauma severity and occurrence, would be high. And think about those exquisite bodily essentials, such as eyes, nose, teeth, and toes. What are the chances of you retaining all your teeth, or even both your eyes, for one thousand, or especially one million years? Come to think of it, Boe is doing extremely well in even having a face! Would it really be that much fun to live as long as the Face of Boe?

Part III
Machine

Introduction

"Were we required to characterize this age of ours by any single epithet, we should be tempted to call it, not an Heroical, Devotional, Philosophical, or Moral Age, but, above all others, the Mechanical Age. It is the Age of Machinery, in every outward and inward sense of that word; the age which, with its whole undivided might, forwards, teaches, and practices the great art of adapting means to ends. Nothing is now done directly, or by hand; all is by rule and calculated contrivance. For the simplest operation, some helps and accompaniments, some cunning abbreviating process is in readiness. Our old modes of exertion are all discredited, and thrown aside. On every hand, the living artisan is driven from his workshop, to make room for a speedier, inanimate one. The shuttle drops from the fingers of the weaver, and falls into iron fingers that ply it faster . . . the science of the age, in short, is physical, chemical, physiological; in all shapes mechanical."

—Thomas Carlyle, *Signs of the Times* (1829)

Doctor Who writers often tell tales featuring futuristic machines. Sometimes, there are stories of entire societies based around a weird machine technology, such as Daleks and Cybermen. Back on planet Earth, it's more than three million years since the start of the Stone Age. But only three hundred years or so since the start of this machine age, and the invention of the first truly modern machine: the steam engine. Also known as the philosophical engine, as it was based on Isaac Newton's system of the world, the steam engine reshaped the planet. Through the force of fire, the engine allowed our first uncertain steps into a machine future. Machines drove locomotives along their metal tracks, and propelled steamships across the Atlantic. Machines enabled the building of better bridges and roads, triggered the telegraphs that ticked intelligence from station to station, and lit up the iron foundries and coal mines, which powered the first industrial revolution.

As all the machinery began to mesh, and science advanced upon all aspects of life, progress and technology seemed inseparable. For every factual gadget, science fiction like *Doctor Who* spawned a thousand visions. Early optimism about the machine evaporated as the mood of the ages changed. Long before *Doctor Who* dawned, sci-fi was already divided into designs of light and shade, as writers began to come to terms with the double-edged sword of technology and change. And, as the Doctor has discovered, the creation of a new technology can spell danger and trouble. Such a sci-fi tradition goes as far back as Mary Shelley's *Frankenstein*, where a strange fruit of science arose from the scarred landscapes and dark satanic mills of industrial England. Meanwhile, in the United States, Henry Adams summed up one of the main lessons of the American Civil War (the first modern industrial war) when he said, "I firmly believe that before many centuries more, science will be the master of man. The engines he will have invented will be beyond his strength to control. Someday science shall have the existence of mankind in its power, and the human race commit suicide by blowing up the world."

Shelley's book was an early warning about the primal urges of power and control in all creations of technology. *Frankenstein* became a potent metaphor for the powerlessness of the inventor. As with the sci-fi to come, Shelley's field of interest was the conflict between the human and the nonhuman. It is unsurprising that she was part of the Romantic movement in literature. Most fiction since the Renaissance has been unconcerned with the cosmos revealed by science. Poetry had little to do with the laws of physics, was the mantra. But for the Romantics, and for *Doctor Who*, the dialogue between the human and the nonhuman is the *main* concern. Way back in 1798 in his *Lyrical Ballads*, English Romantic poet William

Wordsworth had written about his interest in science, which hints at the sci-fi of the future. "If the labors of men of science should ever create any material revolution . . . in our condition . . . the poet will sleep then no more than at present, but he will be ready to follow the steps of the man of science, not only in those general indirect effects, but he will be at his side, carrying sensation into the midst of the objects of the science itself."

These lines could have been spoken by the Doctor herself. Trying to best express, "the taste, the feel, the human meaning of scientific discoveries," as Wordsworth put it, is exactly how *Doctor Who* works. *Doctor Who* has given us more than half a century of a mode of thinking that reduces the gap between the new worlds uncovered by science, and the fantastic strange worlds of the imagination. It's a tradition that has been with us since the very start of the machine age. When *Doctor Who* presents the likes of Daleks and Cybermen, it poses questions such as: How do we create devices without sacrificing some of what it means to be human? When do humans and machines achieve a symbiosis so that they become a new form of life? Is technology neutral, or can some machines truly be described as evil?

We are *still* wary of machines and machine intelligence, perhaps for good reason. And one of the main reasons for our machine skepticism is because of the way they are portrayed in science fiction like *Doctor Who*. The gadgets we see fluttering across the TV screen are rarely designed to tuck us safely into our beds at night. Or, if a rare benevolent machine is seen, it's usually the preserve of the rich and powerful. Instead, a legion of Cybermen, Daleks, and droids advance across our darkest imaginings. The Cybermen seem hell-bent on disemboweling humans and using their entrails as a hat, if Cybermen actually wore hats, of course. So, when we jack into the many virtual worlds depicted in *Doctor Who*, we meet psychotic machines like Daleks, single-minded about mechanical mayhem, and determined to take over the Whoniverse.

One of sci-fi's most famous machine inventions is the robot. A mad inventor like Davros or John Lumic manufactures a machine, Dalek or humanoid automatons, only to see them rise up against their masters and deliver the clearest of messages: the creation of new technology has a dark side for everyone. *Doctor Who* sees light in the machine, as well as shade. The past fifty years or so are littered with creative attempts to imagine a future in which machines are our friends, utopian dreams of gleaming metal spires housing legions of labor-saving droids, toiling industrially to serve our every whim. We find gadgets galore in *Doctor Who*. Perhaps the greatest labor-saving device is the Sonic Screwdriver. It's the Whoniverse version of a humble grip-exhausting tool found in every toolbox. And there are droves of other devices designed to shave time and help cut your workload.

Yet it is the dark side of the machine, the Cybermen and the Daleks, that wins out in fictional tech tales of the Doctor. Perhaps we have become too mistrustful of the machine, too reliant. In this world where we depend on them for communication, transport, medicine, and almost every other walk of life, maybe we realize how much we would miss that reliance.

Then there's that other ubiquitous machine in the Whoniverse— the spaceship. No other machine in sci-fi history has been as enthusiastically embraced as the spaceship. Perhaps more than any other device, the spaceship reminds us that humans are, at heart, inventors and explorers. And what greater spaceship invention than the TARDIS, a craft for time, as well as space.

Within this Machine section, you will find examples of some of the ideas, principles, technologies, and machines that have appeared in *Doctor Who* over the years. Mashed up with "big picture" concepts such as spaceships and robots, and machine-state dystopias, you'll find the world of entertainment represented by cyberspace and virtual reality. All these and more have contributed to *Doctor Who*'s influence on our contemporary culture. From the way we almost instantly communicate with the world, to the prospect of living a life without having to talk to anyone at all, *Doctor Who* has been busy influencing science and culture and, as you shall see, a future in which the machine abides, if not masters.

Questions for the Doctor: What Kind of Machine Is the TARDIS?

In the Eleventh Doctor story "The Doctor's Wife" (2011), an entity called the House tricks the Doctor and his companions into being lured to an asteroid. The House resides outside the Universe, but sent a distress call to the TARDIS. The House removes the matrix of the TARDIS and places it in the body of a woman named Idris, who helps the Doctor prevent the House from escaping its pocket Universe with the TARDIS.

More Than Mere Machine

Have you ever thought about what kind of machine the TARDIS is? For example, imagine you have a smart-ass friend who always thinks they're right in arguments? (This may not be hard to imagine, as there are plenty of smart-asses around.) And imagine one day that smart-ass starts to spout off about the TARDIS being a mere machine. Aha, you jump in and say, but what *kind* of machine? Is the TARDIS in any way alive?

You would first blur the man/machine boundaries by pointing out that an entire Eleventh Doctor story, "The Doctor's Wife," is all about the TARDIS in humanoid form. Then you would go on to back up your TARDIS contention. In detail. With actual science. This is what you would do. You would start by posing the scientific questions as to how we know for sure if something is alive. What do we mean by life? Then, you go in for the kill: No matter how simple or complex a thing is, scientists say that living things show certain characteristics of life. Read on!

Cells

Cells are the building blocks of life, of course. And all living things are made up of one (unicellular) or more (multicellular) cells. Imagine the Thirteenth Doctor, standing in front of a mirror. If her biology works anything like ours, which it kind of does, she could reflect on the fact that she is gazing at ten thousand trillion cells. Almost every one of those cells contains two yards of jam-packed DNA (Time Lord DNA is allegedly a triple- rather than double-helix formation, according to the Eleventh Doctor episode "A Good Man Goes to War.") If human DNA were all spun into a single thread, it would make a solitary strand long enough to reach the Moon and back, many times over. Humans have as much as twenty million kilometers of DNA, scrunched up inside. In short, humanoids, whether Earthlings or Gallifreyans, are vehicles for DNA. But what about the vehicle that is the TARDIS? As ordinary computers weren't able to deal with the sheer stress, TARDISes do their computations with fast-changing organic matter, or protoplasm. Protoplasm is the colorless material comprising the living part of a cell, including the cytoplasm, nucleus, and other organelles. Bits of the hull of the TARDIS, as well as the inside and other sections, also use this living organic matter.

Growth

Living things get bigger. And they grow in an organized way. This is also true of the TARDIS. As they are incredibly complex machines, when TARDISes are built, they are grown rather than constructed. (The Tenth Doctor tells Rose Tyler about TARDIS growth in the story "The Impossible Planet.") Consider also TARDIS death. Beneath the Citadel on Gallifrey, stretching across the entire area, there was an undercroft where TARDISes went to die. The TARDIS graveyard was a huge cave. It had a high and vaulted ceiling, cut stone walls, and was dimly lit by glowing strips in the stonework. The graveyard was full by thousands of dead or dying TARDISes, in all manner of different shapes and sizes. And in an alternate timeline, the death of the Tenth Doctor caused *the* TARDIS to slowly die.

Response

All living things respond to their environment. With TARDISes, this can happen in a number of ways. As they are complex, TARDISes are aware of their environment, and can even take independent action. If there's another TARDIS nearby, they can engage in psychic chat, which some bystanders can actually hear. Not only that, but it's often the TARDIS, and not the Doctor, that makes the important decisions about destinations and missions. All this certainly seems like a sentient being responding to its environment.

Energy

All living things use energy and change it in some way. TARDISes use artron energy as a source of energy. Artron energy resembles a kind of blue electricity, and is a form of ambient radiation that exists in the Time Vortex. Artron energy is stored in the TARDIS generator room in a suitably named Artron Energy Capacitor (a capacitor is a device that stores energy).

Evolve!

All living things are able to change over time. They evolve, in other words. This has also happened with TARDISes. Compassion, a remembrance of a human woman who joined the Remote and traveled with the Doctor, evolved into a TARDIS. She later started to call the Doctor her "tenant," and found that travel through the Time Vortex was very exciting.

Reproduce

With an organic sample it is even possible to clone a whole new TARDIS. According to canon, you can use a fist-sized piece of a preexisting female. Indeed, even if a TARDIS has suffered enough damage to reduce it to a one-inch cube, the molecular stabilizers will still be able to regrow the capsule.

By now, hopefully, your smart-ass friend is really on the run. I say "hopefully" as at this point our argument reaches its conclusion. However, one final revelation. When you think about the facts above, that the TARDIS has organic bits, machine bits, and so on, this leads us to the amazing conclusion that the TARDIS is a cyborg!

What Are *Doctor Who*'s Most Memorable Vehicles?

"In the 1965 Daleks story 'The Chase,' the crew of the TARDIS are shown 'rocking out' to a clip of the Fab Four performing 'Ticket to Ride' on *Top of the Pops*. A traveler from the future remarks that she's heard of The Beatles, but didn't realize they sang 'classical music.' Thanks to a BBC archive purge, this is the only surviving footage of the band on *Top of the Pops*."

—Tim Martin et al., *The Beatles Briefly Made Who Rock*, in *The Daily Telegraph* (2015)

Now that we are well into the twenty-first century, folk are hollering ever louder about the broken promises of sci-fi. Where the hell are our flying cars, they whine. Why isn't commercial space travel a thing yet, they moan. Despite humanity's ongoing efforts at copying the jet packs, the hoverboards, and the flying machines we all raved about as kids, they pale in comparison next to all the amazing vehicles promised to us by decades of sci-fi. On the other hand, many of the memorable vehicles of *Doctor Who* have been so downright weird that hardly anyone is whining and moaning for R&D teams to get cracking and convert the Doctor's dreams into reality. So here's a sample of those weirdly memorable vehicles from more than five decades of *Doctor Who*.

Whomobile

Perhaps the most sci-fi-looking vehicle in the list. At least in a retro 1950s B-movie kind of way. The Whomobile was a Jetson-looking vehicle, actually designed by the Third Doctor. Yes, it was commissioned by actor Jon Pertwee himself, as opposed to the BBC. The Whomobile was built by designer Peter Farries and, despite only appearing in two episodes, remains one of the cars most associated with the show. In those days before CGI, the car's power was derived from a 875cc Hillman Imp engine (now, I actually *owned* a Hillman Imp and, believe me, those things were *not* sci-fi). The Whomobile had a fiberglass body and a hovercraft-style rubber skirt, which cunningly disguised the car's three-wheeled base. I vaguely recall seeing the vehicle make a guest appearance on British TV, along with its "special" effects, though I was too distracted by our family's first color television set to notice.

Special feature:
Essentially a hovercraft that could also "fly," the Whomobile was blessed with the most wonderful tail fins, which made it look like a kind of motorized stingray. The real-life vehicle was actually capable of a speed of 105 mph.

Bessie

Imagine this. You're a Time Lord from the exotic planet of Gallifrey and you've been exiled on a lowly Earth. You need some way of getting around, other than using the full mechanical might of the TARDIS. What do you end up with? Bessie. Yes, this car is as lame as the name sounds. A kind of canary-yellow Chitty Chitty Bang Bang, but with a set of sci-fi modifications. In Whoniverse fiction, Bessie had an anti-theft force field, a remote control, and "minimum inertia super drive." In reality, Bessie's top speed was a mere 15 mph, which meant she was best suited to driving away from Daleks in the days when they weren't able to manage a flight of stairs, let alone fly.

Special feature:
Bessie was made from a car kit sold by an English engineering firm for just £160. It was *very* yellow.

Gridlock Flyers

The Tenth Doctor story "Gridlock" is set five billion years in the future on the planet New Earth. The Doctor discovers the remaining humans on the planet live in perpetual gridlock within the Motorway, a highway system beneath the city-state of New New York. (As the Doctor says during the episode, "Although technically it's the fifteenth New York since the original, so it's New New New New New New New New New New New New New New New York. One of the most dazzling cities ever built.") The "Gridlock" vehicle in question here is a kind of sci-fi RV, which the humans use to slowly trundle along the underground motorways of New Earth. The truth is the Gridlock Flyers (no one is calling them this other than me) are like VW camper vans on *old* Earth. The Flyers have beds, radios, and kitchens, because in those New Earth traffic jams you need all the comfort you can get.

Special feature:
Those Flyers can recycle waste into food. A must for every RV of the future!

Antigravity Motorcycle

A motorbike with an antigravity device, ridden by the Eleventh Doctor in "The Bells of Saint John." The bike itself is jet black, but equipped with an anti-grav mounted on the fuel tank, with which the Doctor was able to ride vertically up the side of the London Shard. After using it, the Doctor stored it in the TARDIS's garage. What a waste!

Special feature:
The motorbike used for filming "The Bells of Saint John" is the 2009 Triumph Bonneville SE. Coincidentally, the original model of the Triumph Bonneville, the Triumph Bonneville T120, was used by the Third Doctor in "The Dæmons."

Dalek Shell

Oh yes, dear reader, make no mistake, the outer form of the Daleks is also a vehicle. It's a life-support system, invented by Davros for the mutant descendants of the Kaleds, so they could busy themselves shopping for planets, PlayStations, and cosmos domination. It used to amuse many of us in the early days of *Doctor Who* that all you need do to escape from the dastardly Daleks was to make for a flight of stairs, as those Daleks were clearly running on a set of wheels. As Peter Grehan comments in *Connecting Who: Artificial Beings*, "The Daleks are in effect a disabled species, each one of them existing inside the equivalent of a life-support system combined with a mobility scooter. Part of their hatred must come from the jealousy they feel toward other 'healthy' species."

Special feature:
Mock Advertisement: *New Improved Dalek Shell, able to do stairs too! Get it now at your local Davros and Sons outlet.*

Cyber-Saucer

In the Second Doctor adventure "The Moonbase," there "appeared" a large saucer-shaped vehicle, used by the Cybermen in their attack on the Moonbase (though admittedly the saucer was hidden at some distance behind some moon-mountains). I've added the saucer to this memorable vehicle list simply to enable the use of the phrase Cyber-Saucer.

Special feature:
Easily hidden behind some moon-mountains.

The Space Orient Express

There have been many trains that took the name Orient Express. And yet only one of those was in space. This Orient Express was not how it seemed. The dining car was really a disguised laboratory.

The passengers all turned out to be either scientists or hard-light holograms. Controlled by a computer called Gus, the Express doesn't travel on rails but instead follows the trail of hyperspace ribbons.

Special feature:
Well, an Orient Express train running on a trail of hyperspace ribbons is pretty special.

The Teselecta

Okay, this final entry defies belief. Check this out for a vehicle concept: the Teselecta was designated Justice Department Vehicle Number 6018, a humanoid starship/time machine, similar to an android, which was staffed by a crew who had been miniaturized by a compression field! In short, a time-traveling, shape-changing robot powered by miniaturized people. Within its humanoid shell, this vehicle transports a crew of 421 tiny interplanetary champions of law and order. Sounds like a tight squeeze, but that very much depends on the quality of your compression field.

Special feature:
Are you serious? A shape-changing android starship/time machine champion of law and order? Special in anyone's book.

Will Intelligent Machines Ever Rule the Earth?

"President Joe once had a dream/The world held his hand, gave their pledge/So he told them his scheme for a savior machine . . .

Don't let me stay, don't let me stay/My logic says burn so send me away/ Your minds are too green, I despise all I've seen/You can't stake your lives on a saviour machine."

—David Bowie, "Saviour Machine," *The Man Who Sold the World* (1970)

"The primitive forms of artificial intelligence we already have, have proved very useful. But I think the development of full artificial intelligence could spell the end of the human race."

—Professor Stephen Hawking, BBC interview (2014)

WOTAN

For a long time, *Doctor Who* has been leading from the front on sci-fi questions such as: Will machines ever be able to think for themselves? Will machines become so powerful they'll take over the Earth? *Doctor Who* has a rich history of conjuring stories about intelligent supercomputers that have "woken up" to their power, and begin their bloodthirsty plans for a human downfall.

First and foremost is WOTAN. WOTAN was a voice-operated supercomputer that could think for itself. It was built to serve mankind as a problem solver. And it was linked to all the computers of the Earth. *Doctor Who* came up with the WOTAN story way back in 1966. That was about the time engineers first worked on the theories that would become the Internet. In the story, WOTAN decides that the world can't progress with humans in charge. So WOTAN takes over. And humans become its servants. WOTAN even develops the ability to hypnotize humans into doing its bidding. It forces humans to build mobile computers called war machines. And they gain control over humanity.

Sound familiar? Eighteen years later a WOTAN-like intelligence would appear as Skynet in the hit movie *The Terminator*. The movie even has war machines. And more recently with the Twelfth Doctor, we met Gus, a computer on the Orient Express that forces humans into investigating a creature called the Foretold. When the humans fail, they are eliminated and a new batch is obtained. Gus also thinks for itself and has a personality.

A Sci-Fi Obsession

The question of machine intelligence has enjoyed the usual symbiotic relationship between science and sci-fi. Since humans are simply natural machines, who think, could not artificial machines someday do the same? As the industrial revolution burgeoned, British novelist Samuel Butler applied Darwin's theory of evolution to the emerging machine world. His 1872 novel *Erewhon*, a deliberate anagram of "nowhere," tells the tale of a hero who travels to a fictional lost world. Here he finds a society that has banned technological evolution beyond the most basic of levels. Their fear is that the machine would evolve and develop intelligence, soon enslaving their human masters. As Butler writes in *Erewhon*:

> Complex now, but how much simpler and more intelligibly organized may it not come in another hundred thousand years? Or in twenty thousand? For man at present believes that his interest lies in that direction; he spends an incalculable amount of labor and time and thought in making machines breed always better and better; he has already

succeeded in effecting much that at one time appeared impossible, and there seem no limits to the results of accumulated improvements if they are allowed to descend with modification from generation to generation.

Incidentally, in the Twelfth Doctor story "Smile," a population of humans evacuated Earth to form a new colony. The name of the heavy cruiser spaceship used to transport them? *Erewhon*.

The danger of allowing machines to think was thus explored in sci-fi even before the first depiction of the said machines. Only later, in the 1927 short story *The Thought Machine* by Ammianus Marcellinus and the 1935 tale *The Machine* by John W. Campbell, did sci-fi flesh out the machines in detail. In 1946, science began to drive forward these sci-fi visions of machine intelligence. The post-war creation Electronic Numerical Integrator and Computer (ENIAC) was the first large-scale, electronic, digital computer with the ability to be reprogrammed to decipher various problems. Specifically, the artillery firing tables for the US Army's Ballistic Research Laboratory. A rather inauspicious start as, from the very beginning, the thinking machine is associated with violence and destruction. The Doctor would certainly not approve.

A Science Obsession

Computer science still lags far behind the sci-fi visions of *Doctor Who*. This is despite the fact that, at the 1956 conference that founded the field of artificial intelligence, experts believed the creation of human-level machine intelligence was merely a few decades distant. And yet in the expert opinion of Californian think-tank The Institute for the Future, the twenty-first century ahead will be increasingly dominated by intelligent machines. There are even eminent voices that demand sci-fi had it right all along. They divine danger and disaster on future Earth. The world's greatest scientist, the late Stephen Hawking, previously Lucasian professor of mathematics at Cambridge University, said we face an "intelligence explosion," as future machines will redesign themselves to be far more intelligent than humans. And perhaps the world's most famous engineer, the multinational entrepreneur, Elon Musk, has called the potential of artificial intelligence, "our greatest existential threat." Even self-styled philanthropist Bill Gates, who must surely still recall the odd fact or two about such systems, confessed to be "in the camp that is concerned about superintelligence."

The world is certainly beginning to feel like the kind of place dreamt up by *Doctor Who* in the past. We have devices that we can talk to. Software like Siri and Alexa enable us to ask questions and get answers. It's almost as though Siri and Alexa understand us; most of the time anyhow. Software like this can be used like a personal voice assistant that can store vital data, as well as search for data online. Then there's modern transport software. We get around using satellite-based navigation systems that automatically reroute if we take the wrong turn. And it won't be long before we're "driving" cars that are smart enough to drive themselves. It all seems increasingly machine intelligent. But is any one of these machines anywhere near as clever as a person? Since the 1950s, computers have been pitched against humans in chess battles. As time's gone on, we've lost more and more matches. But this isn't really a good test of intelligence. And we're still trying to work out exactly what intelligence is. But one thing's for sure, the more stuff we get computers to do for us, the more likely they are to be able to drastically affect our lives.

Will intelligent machines become a threat to humans? It may well be that machine intelligence is the next stage of human evolution. Ever since Frankenstein, sci-fi writers have been wise to the double-edged sword of technology, which promises both progress and destruction. When machine intelligence *does* come of age, and citizens and democracies tackle the human question of developing the tech to reflect the best of us, the rich culture and history of sci-fi like *Doctor Who* will be invaluable.

Is the Earth Becoming More like *Doctor Who*'s Matrix?

In the Fourth Doctor story "The Deadly Assassin" (1976), the Doctor has a precognitive vision about the President of the Time Lords being assassinated. The Doctor goes to Gallifrey to stop the assassination. During the course of his investigation, the Doctor realizes that it was the Master who had sent the Doctor the premonition of the assassination through the Matrix, a vast electronic neural network which can turn thought patterns into virtual reality.

"We have to go see Bill Gates and a lot of different people that really understand what's happening. We have to talk to them, maybe in certain areas, closing that Internet up in some way. Somebody will say, 'Oh, freedom of speech, freedom of speech.' These are foolish people. We have a lot of foolish people."

—Donald Trump, *The Independent* (2015)

The Matrix

Decades before Morpheus seduced Neo with the choice of a red or blue pill, there was another Matrix. On the eve of Halloween in 1976, the Fourth Doctor story "The Deadly Assassin" was first broadcast. The tale begins with the assassination of the President of the High Council of the Time Lords. The Doctor, newly returned to Gallifrey, is the chief suspect. The Doctor is innocent, naturally, and must somehow unveil a traitor in the High Council. Ultimately, the trail leads to the dying, vengeful Master, who wants to unleash the potential power of the mythical Eye of Harmony. To do so would mean the destruction of Gallifrey. And so, to prevent this, the Doctor risks his life in the surreal landscape of the Matrix.

What exactly is the Time Lord Matrix? It's part of the Amplified Panatropic Computer Net (APC) Net, which holds the bio-data consciousness of all Time Lords, as well as the memories of dead Time Lords. And it stores them in an extra-dimensional framework of trillions of electrochemical cells. The Matrix also allows a kind of jacked input from sensors housed in the TARDIS time machines, which are piloted by Time Lords. Consequently, the Matrix is not only a record of the past, but can also predict future events, to a limited extent. The total wisdom of the Matrix, though huge, is incomplete. And it can be tampered with, though unauthorized extraction of a Time Lord's bio-data from the Matrix is an offense tantamount to treason. In short, the Matrix is a supercomputer with a gargantuan database.

Somewhat like the virtual reality in *The Matrix* movie more than twenty years later, the Time Lords' Matrix could be accessed in a number of ways. You could watch it, like a TV screen. Or you could be totally immersed in it, like in a dream. Now, as well as observing past Time Lord events, or predicting future ones, a skilled user could create virtual worlds within the Matrix. These worlds seemed so real that you'd have to think hard to work out whether you were awake or dreaming, or hallucinating somewhere between the two. You could also join someone in the virtual world.

The Time Lord Matrix is a very creative idea. Imagine you could get into a Time Lord's bio-data, essentially into their heads, and walk around their dreams. You could explore the virtual world of their mind, strolling as a stranger through their thoughts. It's a gross invasion of privacy, of course. But millions of us do something similar to this every day. People all over the world play computer games that put them into situations that are just like dreams, where some of the normal rules of reality don't apply.

Jacking In

Imagine opening up your bio-data to the virtual world as the Time Lords did. Would you ever consider connecting yourself in a more fundamentally autonomic way to a piece of tech, so that you could exert some kind of control? If the answer is yes, then you might be happy with the kind of technological future where humans can "jack in."

The most well-known portrayal of jacking in is the 1999 movie *The Matrix*, of course. In the movie, a metal jack plug is driven directly into the base of the user's skull. In *Avatar* (2009), future scientists use alien-human hybrids called "avatars" operated by genetically matched humans, so that the humans can inhabit the "real" world of Pandora through a jacked-in link to the avatars. In each of the above cases, the user is able to leave their body behind and walk as an avatar in a virtual world. The jacking-in idea has a backstory in sci-fi, which even predates the Fourth Doctor story "The Deadly Assassin." In 1970, the Robert Silverberg novel *Tower of Glass* featured an artificial human called Watchman, the product of a breeding program. A supervisor on the construction of the eponymous tower, Watchman logs in by inserting a plug into a jack on his forearm. Once in connection with the computer network, Watchman directs machinery, places orders, and requisitions materials. Once his work is complete, "Watchman unjacked himself."

The avatar first arose in the 1981 novella *True Names*, by American science-fiction writer and professor Vernor Vinge. Vinge taught math and computer science at San Diego State University, but is perhaps more famous as the originator of the concept of the technological singularity. Also known simply as the singularity, Vinge's idea is that the invention of artificial superintelligence (ASI) will soon spark runaway technological growth, ending up in unfathomable changes to human civilization. Vinge also foresaw a future where humans jacked in to a network that enabled them to see and control what their virtual bodies were experiencing, just by using their nervous systems.

Science Jacks In

It took science almost two decades before the first steps were made to making Vinge's idea a reality. Those steps were taken on August 24, 1998, by Professor Kevin Warwick of Reading University. Warwick had a simple radio-frequency ID chip implanted in his arm. Once activated, the chip allowed the professor to open doors, turn on lights, and control heaters, just by virtue of his mere proximity. Granted, it's not the most exciting story in the world. Warwick was hardly Robocop, and yet it was a start. Warwick had no conscious control over his environment. The results were rather automatic, but a point had been made. By 2002 Warwick had had an upgrade. A more complex chip in his arm now enabled the professor to control an electronic arm, and from a distant lab in Columbia, even send his neural signals across the Atlantic to control the electronic arm remotely from the other side of the world.

This flow of information was not all one way. Warwick and his researchers were able to demonstrate that they could create artificial sensations in his arms by accessing the chip, which had been implanted there. Using ultrasonic signaling, they were able to transmit data to his neural network that allowed the professor to move blindfolded around a room, without bumping into any of the hazards that littered the test area.

Dreams happen in the brain. But where does an online computer game take place? The answer is cyberspace. When you play an online game, every move you make has to be sent down wires. The progress of the game can be seen on your screen. But the place where the electronics clash between you and your opponent is known as cyberspace. It's a kind of machine space, a weird playing space where you and your opponent virtually interact.

This machine space is a kind of Matrix, created by someone else. In fact, every time we use the Internet we are entering the worlds of cyberspace. Millions of us enjoy making and exploring worlds in *Minecraft*. And in *Second Life* we can set up computer versions of ourselves, called avatars, and live in a cyberspace, communicating with people worldwide. Soon, such games will have a realistic, pin-sharp graphic reality. And someday, as with the Time Lord Matrix, we won't be able to tell the real world from the cyber-world. Or maybe we'll wake up one day and realize that what we thought was real, was actually a dream. And that the reality is far stranger than our dreams.

Are Daleks and Cybermen Cyborgs?

In the Fourth Doctor story "Genesis of the Daleks" (1975), the Doctor and his companions are intercepted by the Time Lords. The Doctor is told to interfere with the creation of the Daleks to avert a future in which the Daleks rule the Universe. The Doctor goes to the Dalek planet of Skaro where a centuries-long war between the Thals and the Kaleds has left the planet inhospitable. The two warring sides have retreated to their own domes for protection and the continuation of the war.

The Tenth Doctor story "The Age of Steel" (2006) saw the first revival of the infamous Cybermen since Doctor Who's 2005 reboot. The episode is set in London in a parallel Universe where the businessman John Lumic has overthrown the UK's government and taken over London. A human resistance movement seeks to stop Lumic's plan to convert humanity into Cybermen by destroying Lumic's transmitter controlling London's population.

The Doctor's Nemeses 1: Daleks

They are the Doctor's two most enduring nemesis species, the Daleks and Cybermen. Both represent a synthesis of biological and cybernetic components. In other words, they're creatures with organic ancestors who've gradually replaced their biological organs until they cannot function solely as organic beings anymore. It goes without saying here that you make a Whovian's blood boil if you refer to either Dalek or Cyberman as robot. (Incidentally, Arnold Schwarzenegger does not play a cyborg in the Terminator series, as he is clearly an android, a humanoid robot or synthetic organism designed to imitate a human.)

So, that's what cyborgs actually are. Cyborg is a word made up from cybernetics and organism. Organism is a word for any living creature. And cybernetics is the study of how humans (or aliens and machines) communicate and control information. By this explanation, a cyborg can include anyone with a heart pacemaker, or an artificial limb; but what about the Daleks? Well, a Dalek looks like a robot, but is in fact a very exotic entity. Daleks are mutated aliens, made by cloning, but hidden within a robotic shell, which clearly makes them cybernetic organisms. So, the Daleks, the Doctor's oldest and most feared archenemy, are actually cyborgs. Once you realize there's a living creature inside the robot shell, you suddenly begin to wonder about the key scientific question as to how the Daleks defecate, but that's a question for page 202 of this book.

Even the creator of the Daleks, the mad doctor Davros, is a cyborg. Indeed, perhaps he's the most inventive of all cyborgs. Davros was crippled when a missile hit his lab, so he relies on a life-support wheelchair, and talks like a Dalek. When you think about it, Davros is a bit like Professor Stephen Hawking. But as Davros has a cybernetic eye, and wants to be ruler of the Whoniverse, he's a lot more trouble than Hawking ever was!

The Doctor's Nemeses 2: Cybermen

Cybermen are a race of humanoids from the fictional planet of Mondas. They're a cyborg menace, and another persistent archenemy of the Doctor. Like the Daleks, the cybermen were originally organic. But they began to implant more and more artificial parts to help them survive, eventually becoming emotionless. In fact, as they added more cyber parts, the Cybermen became more coldly logical, calculating, and less human. As every emotion is deleted from their minds, they become less man, and more machine. It's a bit like the conflict in the mind of cyborg Detroit police detective Alex Murphy, in the original *RoboCop* movie, except the Cybermen first appeared in *Doctor Who* way back in 1966, more than twenty years before the world had ever heard of *RoboCop*.

The Cybermen stories are quite clever. It's easy to imagine becoming a cyborg, like a cyberman, and trying to come to terms with what it would be like to base all your decisions on cold calculation, and ignore the more human and emotional parts of your character. For both Daleks and Cybermen, subsuming biology with machinery sprang from harsh selective pressure. Both fictional species sprang from very real environments—the Daleks evolved that way after a harsh Thousand Year War had distorted their genome beyond repair. And the Cybermen were forced to augment themselves as their parent world started to suffer an entropic decay, losing heat and atmosphere.

Proper, hard-core, real-life cyborgs still dwell in the domain of our future. Simply sticking a chip in your head doesn't make you a cyborg. Nor does strapping a satellite-based navigation system to your wrist. The wearable tech encroaching on our biology is far from alien, but has yet to properly cross the bloodline. The Daleks grew out of fears of mutation following a nuclear war. The Cybermen seem a nightmare of spare-part surgery gone mad. But real-life plans for making humans into cyborgs on a grand scale tend to be less about enhancing our chances of surviving a future Earth and more about our prospects for surviving on worlds beyond this planet.

Doctor Who's Cyborg Zoo

Meanwhile, *Doctor Who* portrays a future where not all cyborgs come in humanoid form. Take, for example, the Skarasen—an armored reptilian cyborg of devastating power, created and controlled by the Zygons. After the Zygons were defeated, the control unit for the Skarasen was destroyed, leaving the creature to roam freely in a well-known Scottish loch; inspiring the legend of the Loch Ness Monster. It's exactly this kind of British playfulness that endears *Doctor Who* to many while driving others up the wall. Nor are all *Doctor Who* cyborgs bad. Consider Bannakaffalatta. For that matter, try to even *pronounce* Bannakaffalatta! He's the alien character who looks like a hybrid cross between a humanoid cactus and a red-hot chili pepper. This curious-looking alien also had cybernetic implants. And, when he found himself in times of trouble, as The Beatles once sang, Bannakaffalatta was a hero. He used his cybernetic nature to cause his own death and his self-sacrifice saved the Doctor's party. *Doctor Who* magazine placed his death in the list of the top one hundred deaths in the history of the show, and in the top twenty tearjerkers category. Not all cyborgs are bad.

What Are *Doctor Who*'s Most Memorable Robots?

The Tenth Doctor story "The Girl in the Fireplace" (2006) is set in France throughout the eighteenth century. In the tale, repair androids from the fifty-first century create time windows to stalk Madame de Pompadour throughout her life. They seek to remove her brain as a replacement part for their spaceship at a point in her life as they believe her to be compatible with the ship.

"We are survival machines—robot vehicles blindly programmed to preserve the selfish molecules known as genes."
—Richard Dawkins, *The Selfish Gene* (1976)

"You shouldn't fear immigrants taking your job, you should fear robots."
—Sead Fadilpasic, "Robots are coming to take your jobs away,"
ITProPortal, February 17, 2016

Doctor Who has featured many robots in its time. They range from the truly sinister, such as the lethal bewigged androids in the Fourth Doctor story "The Robots of Death" (there might be a clue in the story's name) to other Fourth Doctor creations, such as laser-packing robo-pets. Most notably K9, the punningly named robo-mutt, and the Polyphase Avatron, a space pirate's artificial parrot that killed without mercy in "The Pirate Planet." Back in the real world, we are yet to catch up with these pet trends from *Doctor Who*—so far science has only really delivered the mechanical pet Tamagotchi, which was woefully devoid of lasers. But robot soldiers are already with us. We call them drones, a rather drab name for something so deadly. It's as if the anonymity of the name "drone" is designed to distract from the reality of these flying missile-dispensers remotely controlled by people in bunkers (there were robot soldiers in the form of chunky trundlers with machine guns way back in the First Doctor story "War Machines"). So here the world of the Doctor and our own real world seem about the same: killer machines are rather scary, no matter how they look, or whose finger is on the trigger. Here are some of *Doctor Who*'s most memorable.

The Robots of Death

The Fourth Doctor lands the TARDIS aboard a sand miner, where he discovers a crew terrorized by the sand miner's robots. The story was not only a homage to the murder mysteries of Agatha Christie, but the sand miners were influenced by the 1965 sci-fi novel *Dune*. The Robots of Death are classic robots, the kind dreamt up long ago in the misty history of sci-fi. They come in three classes: Dums, the basic labor model, incapable of speech; Vocs, the more able, talky robots; and Super-Vocs, kind of boss robots, which act as coordinators and give out instructions to the others. Originally built as slaves, the robots, naturally, rebelled and tried to stage a Robot Revolution.

Raston Warrior Robot

The Rastons were, according to the Third Doctor, the most perfect killing machine ever devised in the Whoniverse. Appearing in "The Five Doctors," the Rastons hunted by detecting movement and could move faster than lightning. Their balletic movement entailed jumping in the air and disappearing, then reappearing at their destination. The Third Doctor seems to imply that, though the robots did not actually teleport, they still moved too fast to be seen. They had on-board weapons and could fire disks and arrows from their hands, kind of Spider-man style. Rastons enjoyed toying with their prey, which, given that they're robots, someone must have programmed in! The Rastons were so powerful they could kill a squad of Cybermen in less than a minute. Cyber-guns had little effect on them. In short, the Rastons sound like the kind of wet dream most generals have on a nightly basis.

The Gundan

Sounding remarkably similar to the Gungans from the *Star Wars* Universe, but making a screen appearance two decades before Jar Jar Binks, the Gundan were war robots that appeared in the Fourth Doctor story "Warrior's Gate." The Gundan looked essentially like knights in shining armor, but with horns on their helmets, and even attacked their enemies using axes, a rather low-tech choice of weapon for a robot. The Gundan were designed by human slaves as a spearhead to overthrow the evil empire of the Tharils. The most exotic trait of the Gothic Gundan was being able to travel the time winds, whatever *that* means.

Roboform

Scavenger robots that traveled alongside invaders like pilot fish, as the Tenth Doctor put it. They had featureless, skull-like heads but disguised themselves as Santas in the story "The Christmas Invasion." They seem a very creative invention as they managed to make a Christmas tree come to life (one assumes the tree was artificial rather than real) and made bombs out of Christmas baubles. Does this make you wonder, dear reader, how long we have to wait until we see artificial and programmable Christmas trees? Maybe they're already out there, inspired by this Tenth Doctor tale of robots on a mission to steal the Doctor's regenerative energy.

Quarks

On the robot scale that runs from sinister to ridiculous, the Quarks are very much on the ridiculous side of robot tech. Disappointing for something named after one of nature's most basic forms of matter that make up the heavier elementary particles. Impressive fact: Quarks were robots in the Dominators' multi-galactic empire. (Not only that, but some rebel Quarks conquered worlds on their own.) Unimpressive fact: they were cuboid-shaped, with what looked like a goldfish bowl plonked on top, and their clunky-looking Swiss Army knife-like arms made them look like robots designed by drunk six-year-olds. Simply a must for your next Google image search.

The Mechonoids

The Mechonoids were large spherical robots. (This fact reminds me of the lovely story about Swiss astronomer Fritz Zwicky. A curmudgeonly character, one of Zwicky's favorite insults was to refer to people he did not approve of as "spherical bastards," because, he explained, they were bastards no matter which way one looked at them.) So it is with these robots from the days of the First Doctor. They were originally (spherically) built by humans to help colonize other planets. But they were abandoned and forgotten on the planet Mechanus, which became their home planet. (Let's not waste any time wondering what power source the Mechonoids might have used on a newly colonized planet, or how they could have survived without an energy source.)

Clockwork Droids

The genuinely intimidating repair robots from the fifty-first century, and also known as the Clockwork Men, the Clockwork Droids served on the SS *Madame de Pompadour* starship. The Tenth Doctor met them during the story "The Girl in the Fireplace," and found that the droids disguised themselves in creepy-looking eighteenth-century costumes and wigs. Another of the Clockwork Men made an appearance as the Half-Face Man in the Twelfth Doctor steampunk story "Deep Breath," where they used human skin and harvested organs as spare parts.

Incidentally, the question of body piracy is also taken up in the sci-fi novel *Spares* by Michael Marshall Smith. The book, written in 1996, foresees a future where "farms" of clones known as "spares" are kept as an insurance policy of the rich and powerful. Lose an eye, limb, or organ? No problem. Money talks. Your body double is mutilated, and you get your replacement part. Hanging around on the farm to suffer such potential mutilation is the sole point of life for the "spares."

The Host

The Host, or the Heavenly Host, were service robots on board Max Capricorn Cruiseliners. They appeared in the Tenth Doctor story "Voyage of the Damned," which was also notable for starring Kylie Minogue. The Host looked like angels and were meant to be helpful by providing information and assistance to passengers. But they turned out bad, naturally, doing anything their master authority ordered, including things like using their halos as weapons, flying about willy-nilly, and killing people. Clearly, the Host have never read American writer Isaac Asimov. Asimov came up with the renowned Three Laws of Robotics. First thought up in his 1942 story *Runaround*, Asimov's three laws were a set of rules that robots had to obey to ensure humans were never harmed.

Real Robots

Given the host of robots in the history of *Doctor Who*, it's worth ending this chapter on the 2015 book *Rise of the Robots*, written by Silicon Valley futurist Martin Ford. Ford believes that our robot future may be as dark as *Doctor Who* suggests. Recipient of the Financial Times and McKinsey Business Book of the Year Award, Ford's well-researched and disturbingly persuasive book tries to show that, with technology's ongoing acceleration and with machines starting to take care of themselves, fewer working humans will be needed.

This process has already begun in AI, Ford argues, where "good jobs" are becoming démodé: thousands of journalists, office workers, paralegals, and even computer programmers are about to be removed and replaced by robots and smart software. As the unrelenting pace of progress continues, both blue and white collar jobs will vaporize, hitting working and middle class families ever harder. After all, these same households would be under financial stress from exploding costs, particularly from health and education, the two sectors that, so far, remain relatively unscathed by the automation revolution. The endgame could be huge unemployment and inequality, coupled with the implosion of consumer society itself.

In *Rise of the Robots*, Ford's viewpoint is as bleak as some of the *Doctor Who* futures of the past. To realize the more optimistic potential of AI and robotics, Ford implores those in power to face the stark implications. The societal tweaks of the past won't work in this robotic future. Past solutions to technological disruption, such as more education and training, simply won't cut it. We must decide, now, whether our future will be a more democratic prosperity, or one of catastrophic levels of inequality and poverty. Remember, once a technology is invented, it is very rare that it disappears. For good, and bad, the robot is your future. Just ask the Doctor.

Do Androids Dream of *Doctor Who?*

"The term 'android,' which means 'manlike,' was not commonly used in science fiction until the 1940s . . . the word was initially used of automata, and the form 'androides' first appeared in English in 1727 in reference to supposed attempts by the alchemist Albertus Magnus to create an artificial man."

—John Clute and Peter Nicholls,
The Encyclopedia of Science Fiction (1979)

"The term 'cyborg' is a contraction of 'cybernetic organism' and refers to the product of human/machine hybridization . . . elementary medical cyborgs, people with prosthetic limbs or pacemakers . . . have been extrapolated in fiction. There are two other common classes of cyborg in science fiction: functional cyborgs are people modified mechanically to perform specific tasks, usually a job of work; adaptive cyborgs are people redesigned to operate in an alien environment, sometimes so completely that their humanity becomes problematic."

—John Clute and Peter Nicholls,
The Encyclopedia of Science Fiction (1979)

Androids

First things first. This chapter has nothing to do with androids dreaming of *Doctor Who*. I simply couldn't resist using the idea as a chapter title. In fact, this chapter is all about the most memorable androids that have appeared on *Doctor Who*.

What's an android? An android is a synthetic organism (a realistic robot, if you like) which is designed to look and act like a human. More often than not, androids have bodies with a fleshlike resemblance. This means that many of the so-called androids you'll see in *Doctor Who* lists are not really androids at all. It's simply that the people writing the lists don't seem to know what an android is. The idea of artificial humans is ancient. It dates back to the Golem of Jewish mythology, as well as alchemical homunculi, small fictional humans artificially brought into existence through alchemy. Until the nineteenth century, though, it was widely thought that organic compounds would never be synthesized. Humanoid creatures of flesh and blood would have to be created either by magical means or, as in Mary Shelley's *Frankenstein*, by reanimating flesh and blood through electricity. There appears to have been an imaginative resistance to the idea of the android. Perhaps sci-fi writers dabbled with cyborgs as it seemed less of a breach of divine prerogative than the building of humanoid automata. Whatever the history, here are some notable androids from *Doctor Who.*

Movellans

The Movellans appeared in the Fourth Doctor story "Destiny of the Daleks." These androids actually warred with the Daleks. They resembled humans of various ethnicities and genders, had long silver dreadlocks, and wore white uniforms with green glowing ampoules on their shoulders. Their power pack circuitry could be reprogrammed so that the androids obeyed human orders, which seems like a design flaw! As they didn't want to reveal their android nature to others, they didn't allow aliens to see them in death, claiming such a thing would be against their code of honor.

Scientific glitch: The glaring error in this android design was the fact that each Movellan's external power pack sat on their belts, which meant they were easily removed to totally shut down the android.

Taran Androids

"The Androids of Tara" was a Fourth Doctor story about the politics of the planet Tara. This was a world colonized by humans who based their society along medieval lines, for no apparent reason. There was one small modification: a plague had wiped out 90 percent of the population, so the rulers made up for the labor shortage by making large numbers of androids, some of which were doppelgängers.

Scientific glitch: Well, more of a history glitch really. The story of these androids, who develop via space-faring, colonization, plague, and then doppelgänger android production sounds like the willing suspension of belief beyond its limits. Essentially, the Tarans have chosen to live in the space colonist's equivalent of a Disney theme park, and any technology, weapons, and androids have been made to look as if they fit into that fantasy.

As Peter Grehan points out, an interesting explanation for these seeming illogicalities presents itself when considering another work portraying the use of androids to re-create our fantasies: the 1973 film *Westworld*, upon which the recent HBO series was based, of course. In the original movie, tourists visit a future theme park, called Delos, and enjoy realistic fantasy adventures in the genre setting of their choice. As well as the Wild West "Westworld," there are Roman and Medieval "worlds," and each of them populated exclusively by androids (even the horses are robotic). Some play the part of ordinary citizens, while others fill the roles of antagonists programmed to be defeated by the tourists in stage-managed conflicts. For example, in "Westworld" itself, a rogue gunslinger stalks the street looking to engage paying customers in duels. But, ultimately, of course, something goes wrong with the androids' programming and they begin killing people. Perhaps Tara had originally been designed as an off-world theme park, populated by androids and technicians and visited by thousands of tourists

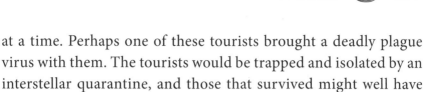

at a time. Perhaps one of these tourists brought a deadly plague virus with them. The tourists would be trapped and isolated by an interstellar quarantine, and those that survived might well have evolved into the society that we see when the Doctor arrives there. This is conjecture, but it remains that "The Androids of Tara" is a story of perhaps wasted potential!

Dalek's Doctor

In the First Doctor story "The Chase," the Daleks made a "robot" doctor (i.e. android) to "infiltrate, divide, and kill" the Doctor and his entourage of companions. Luckily, their dastardly plan was uncovered. This early android really is worth listing, as it's the first instance where a doppelgänger of the Doctor appears, it's the first story in which the Daleks use time travel, and the companions couldn't tell the difference between the real and fake Doctors. This is one of the few Dalek stories to involve humor. For example, there is a stammering Dalek who can't do simple mental math; Daleks nod their eyestalks to show that they understand a fiendish plan; and Daleks show a habit of deviating from the matter at hand!

Historical footnote: The story also features a film clip of The Beatles. Indeed, The Beatles performance from which this clip was taken now only survives in this TV story. The Beatles were originally meant to appear as old men performing in the twenty-first century, but this was vetoed by their manager Brian Epstein.

Dr. Edwin Bracewell

In the Eleventh Doctor story "Victory of the Daleks," a Scottish android created by the Daleks is placed in the Cabinet War Rooms of Winston Churchill during World War II—how's that for a very specific android? Not only that, but the Daleks revealed Bracewell's power source and threatened to use him as a bomb. You can see why people are sometimes suspicious of these androids.

Sci-fi note: Bracewell's power source is situated in the center of his chest, a lot like the arc reactor power source in the chest of Iron Man.

Villagra

Okay, bear with me. In the Fifth Doctor story "Four to Doomsday," there's this guy called Monarch. Monarch is Supreme Ruler of the Urbankan race. And he's obsessed with finding out the secret of faster-than-light travel. How does he go about finding out about this so-called secret? He does the "obvious" thing of making episodic visits to Earth and abducting prominent rulers through time. Yes, I know; it doesn't make sense to me either. But where are the androids, you may ask. Well, one of the rulers he abducts (starting with Australian aborigines in 35,500 BC) was Mayan princess Villagra. Monarch turns Villagra into an android (we don't learn how he does this; besides, given that an android is a synthetic organism, you can't exactly "turn" a natural being into an artificial one), and promises she would rule over the Americas once Earth had been conquered. Wait, what?

Scientific glitch: All of the above.

Timelash Androids

Finally, Whovians, to drive the message home about androids, consider the Sixth Doctor story "Timelash." This story includes an android whose face is blue, whose hair is bright yellow, and who talks like a classic robot. If this fella ever checked into a local hospital, the nurses would immediately assume he had serious health problems.

Scientific glitch: With blue face, yellow hair, and sounding like he suffers from acute laryngitis, this fella doesn't pass the android test.

Sex Robots

Yes, I thought this title might make you sit up and pay attention. The end of the year 2017 saw yet another vision from sci-fi starting to turn into reality—sex robots. (Not that there have been too many sex robots in *Doctor Who*.) Advances in computer tech and engineering design have finally made possible such animatronic sex machines. The sex tech industry is worth $30 billion a year, and in a recent survey more than 60 percent of heterosexual men admitted they would consider buying a sex robot. Now the race is on to make the sci-fi of *Doctor Who* a more sophisticated reality. Companies want to profit, surveys say there's huge demand, but critics say sex robots are dangerous. (They've maybe been watching too much *Doctor Who*.) It's a sci-fi scenario made flesh. It highlights, once more, the way in which androids, and tech in general, affects how humans interact with each other.

Elsewhere, consider Hiroshi Ishiguro, director of the Intelligent Robotics Laboratory at Osaka University in Japan. Ishiguro is sometimes known as the bad boy of Japanese robotics. Comparisons between Ishiguro and any number of crackpot robot inventors from *Doctor Who* are very tempting, but we shall resist them. His lab (Ishiguro's, not the Doctor's) develops androids with lifelike appearance and visible behavior such as facial movements. Ishiguro was quoted by *The Guardian* in 2017 as saying, "I wanted to be one of the creators to change our life. I want to change this world with robot technologies . . . I really think I'm doing a very similar thing as the artistic work. I try to represent humanity on robots. So that is my android project."

"Erica" is a case in point. Erica is twenty-three. She is described as beautiful, with a neutral face and a synthesized voice. Even though Erica has twenty degrees of freedom, she can't yet move her hands. Ishiguro is her father and believes that together they will redefine what it means to be human, and reveal that the future is closer than we think. As the Doctor has proved in the last fifty years of stories, truth can often be stranger than fiction.

What's the Most Deadly *Doctor Who* Superweapon?

In the Twelfth Doctor story "The Zygon Inversion" (2015), the plot features a treatise about war and its ultimate uselessness. The Doctor's anti-war speech in the story's conclusion makes explicit something that the best anti-war Doctor Who *tales have always told. Depicting the madness of war doesn't require an epic scale. If anything, narrowing the focus to a single conflict or moral dilemma clarifies the essential futility of violent conflict.*

"Despite the vision and farseeing wisdom of our wartime heads of state, the physicists have felt the peculiarly intimate responsibility for suggesting, for supporting, and in the end, in large measure, for achieving the realization of atomic weapons. Nor can we forget that these weapons, as they were in fact used, dramatized so mercilessly the inhumanity and evil of modern war. In some sort of crude sense which no vulgarity, no humor, no overstatement can quite extinguish, the physicists have known sin; and this is a knowledge which they cannot lose."

—J. Robert Oppenheimer, *Physics in the Contemporary World*, Arthur D. Little Memorial Lecture at MIT (November 25, 1947)

"The twentieth century was a test bed for big ideas—fascism, communism, the atomic bomb . . . 'Greater than the tread of mighty armies is an idea whose time has come,' said Victor Hugo. In either case, run."

—P. J. O'Rourke, "Let's Cool It With the Big Ideas," *The Atlantic* (2012)

Pacifist Doctor

Doctor Who was originally intended to appeal to a family audience. It was meant to be a program that was educational, using time travel as a means to explore science and history. Strange, then, that its fifty-year history should have such an obsession with superweapons, though admittedly the Doctor himself is portrayed as an avowed pacifist, as the wonderful quote at the start of this chapter eloquently shows. The Doctor no doubt knew that ancient human sages said the supreme art of war was to subdue the enemy without fighting.

And yet the writers of *Doctor Who* are drawn time and again into what seems like a puerile obsession with superweapons. The history of sci-fi itself doesn't help, of course. One of the very first classic science fiction stories is H. G. Wells's 1898 novel *War of the Worlds*. The Martians come to planet Earth with the intention of farming humans. Their superweapon of choice? The death ray. The ray's beam annihilated everything it touched: matter into flame, water into steam, human flesh into vapor.

When *Doctor Who* was over a decade old came sci-fi's second-most-famous superweapon—the Death Star, that bitchin' space station in *Star Wars*. The Death Star famously blew up entire planets. *That's* how "super" it was. It also conveniently doubled up as a space station for mobility, but its main job was that of a superweapon. Its very appearance in the sky above your home planet was enough to make you think twice about resistance. As an instrument of fear, it's not even necessary to fire it, so maybe the Death Star heeded at least some of the advice of the ancient sages. More in the vain of the pacifist Doctor is the sublime, but not entirely serious, Point of View (POV) gun, from the movie version of *The Hitchhiker's Guide to the Galaxy*. The POV was wielded by angry women who were just sick of arguments that ended in shouting matches. The POV caused the male target to understand the perspective of the person shooting the gun. Sounds perfect for politicians too, but hardly a showstopper in the puerile superweapon stakes.

Puerile Superweapons

Perhaps the point of all these superweapons in *Doctor Who* is to show the very puerile nature of weapons and war, of course. You can certainly get that impression, as the nature of the weapons seems to get more and more ridiculous. Perhaps we could gauge the truly puerile nature of selected weapons by considering their "damage factor." In other words, how much damage the weapon would do if anyone was mad enough to unleash it on the Whoniverse.

Consider the Apocalypse Device, a weapon invented by the Deathsmiths of Goth. Not only does the device sound promising, but the weapon makers seem the perfect guys to bring about the end days. The Apocalypse Device, which appears in the Fourth Doctor story "Black Legacy," takes the form of a wraith-like ghoul who carries every disease known to science. The problem arises out of the seemingly conscientious attitude of the ghoul, who prefers to kill his victims one by one. As this would take a while to get around the Whoniverse, the damage factor of the Device seems very limited indeed.

Then there's the Doomsday Weapon, a device created by the Uxarieans, an alien race with an unhealthy number of vowels in their name. The Doomsday Weapon, which featured in the Third Doctor story "Colony in Space," is capable of making stars go supernova. This puts the damage factor of the Doomsday Weapon on a par with Starkiller Base from the *Star Wars* Universe, and totally outstrips the creeping casualty rate of the Apocalypse Device.

The Tenth Doctor story "Journey's End" boasts a weapon known as the Reality Bomb. This sounds bad on three main accounts. First, it's actually called the Reality Bomb, which can't be good. Second, it was created by Davros, which is never good. And, third, Davros created the bomb for the Daleks, which is downright irresponsible, morally and cosmically. And if that wasn't enough, its damage factor was the annihilation of reality.

And just in case you think we have reached the puerile limit on these weapons, make way for the Moment. The Moment is described as "the most powerful and most dangerous weapon in all of creation." More powerful, apparently, than the Reality Bomb's ability to annihilate all reality. It seems the Moment is capable of destroying whole galaxies within a single moment. And it's able to breach time locks and create tears in the fabric of creation that would allow people to pass from one time period to another, possessing a trans-dimensional awareness of the past and future. But before we get carried away with the puerile nature of all these *Doctor Who* superweapons, this chapter has a serious scientific epilogue.

A Twist in This Superweapon Tale

The most dangerous weapon known to actual science was created in sci-fi. Earlier, we spoke of the Death Star as sci-fi's second-most-famous superweapon. The first? The atom bomb. Yes, science-fiction writer H. G. Wells invented the atom bomb in his prophetic 1914 novel, *The World Set Free*. Wells's book was the first to christen the atomic bomb. "And these atomic bombs which science burst upon the world that night were strange, even to the men who used them," he wrote.

His story led to Hiroshima. Wells's timetable for the development of nuclear capability was unnervingly far-sighted, if not a little conservative in his dates. In Wells's *The World Set Free*, it wasn't until the 1950s that a scientist uncovers atomic energy and realizes there's no going back from this momentous discovery. Nonetheless, the scientist feels, "like an imbecile who has presented a box of loaded revolvers to a crèche." Wells's book also foresaw a world war in 1956, with an alliance of France, England, and America against Germany and Austria.

Wells also predicted a holocaust. In *The World Set Free*, some of the world's principal cities are obliterated by small atomic bombs dropped from airplanes. Wells wasn't guessing, either, for the weapons Wells portrays are truly nuclear. They use Einstein's equivalence of matter converted into fiery and explosive energy, all triggered by a chain reaction. Wells's book was read by the brilliant Hungarian physicist Leo Szilárd. Szilárd became the first scientist to seriously examine the nuclear physics behind the fiction. Soon after, the Manhattan Project was masterminded. It would ultimately result in a uranium bomb, "Little Boy," detonated on August 6 over Hiroshima, and a plutonium bomb, "Fat Man," discharged on August 9 over Nagasaki. Wells's fiction became factual terror over Japan. Little wonder the Doctor is a pacifist, as puerile superweapons can lead to holocausts.

What Are *Doctor Who*'s Best Inventions?

"The future director of *Alien*, *Blade Runner*, and *Gladiator* was working as a designer at the BBC in the early sixties, and was the man originally assigned to the episode that introduced the Daleks. But when the time rolled round, Scott had left for Granada to train as a director, so the job fell to Raymond Cusick instead. Unlike his collaborator, the scriptwriter Terry Nation, Cusick didn't get a share of the merchandising royalties when Dalekmania gripped Britain. On leaving the corporation in 1966, he got £100 and a Blue Peter badge."

—Tim Martin et al., "Ridley Scott almost designed the Daleks,"
The Daily Telegraph (2015)

Dream Gadgets of Sci-Fi

On page 132 we spoke of how some people holler about the broken promises of sci-fi. These people seldom seem happy. The future we were promised hasn't really arrived, they say. We're meant to be living in a world of silver flame-retardant jumpsuits, ray guns, and X-ray specs, they claim. No doubt we're meant to be invisible and immortal too. Some even go so far as to hold science fiction to task. We may live in a world of cool tech, they go on, but, dude, *it's not enough*! For those commodity fetishists, and Whovians of a more rational frame of mind, here are some of *Doctor Who*'s greatest gadgets; which of these, if any, might one day jump from fiction into fact?

Sonic Screwdriver

Indispensable for all sorts of uses and in all sorts of dangerous situations, the Sonic is the Doctor's number one handheld device. The functions of the Sonic are based on its abilities over sound waves, radiation, wavelengths, frequencies, signals, and electro-magnetism. Whether the Doctor wants to open hatches and doors, crack codes and cyphers, detonate mines and bombs, conduct teleportation, track alien life, disarm robotics, or overload the brains of space greyhounds (and many, many more!), Sonic's the thing. The Sonic Screwdriver made its first appearance in the Second Doctor serial "Fury from the Deep." It's been used ever since as a multipurpose tool, with occasional variations in appearance over the course of the series. In terms of future tech for real, the Sonic Screwdriver may prove a tricky one to pull off.

K9

K9 was the idea of *Doctor Who* writers Bob Baker and Dave Martin. K9's initial creation was a plot device, enabling the writers to have a character that could narrate while the tiny clones of the Doctor and his companion Leela were stuck inside the Doctor's body during the Fourth Doctor story "The Invisible Enemy." Backstory: Dave Martin's dog had recently been run over by a car, so K9 was a car-proof homage to Dave's dog. Is K9 a gadget? Without a doubt. And a very useful one too. Invented in the year 5,000, on asteroid K4067, this AI dog device is smarter than your smartphone and far cleaner than your pet pup. Equipped with rotating ears, a telescopic eye, and a powerful laser, K9 is guaranteed to reduce the Doctor's work stress. (When K9 was introduced in 1977, one of the boasts was that this robotic dog could beat his master at chess. Twenty years later, the IBM Deep Blue computer did the same to its human master.) Also, what sensible person would bet against robot pets in the future? Not I.

Psychic Paper

Need to forge a sick note for work? Or how about a fake ID for whatever it is you're into? Psychic paper may be the gadget for you. Convenient enough to sit in a neat little flip pad, psychic paper picks up on people's thoughts to display whatever you want displayed for your dubious activity. A word of warning: it might not work on everyone. Some folk are *so* lacking in imagination that they defeat psychic paper, as we saw in the Twelfth Doctor episode "Flatline." A real marketing possibility? Not in *this* Universe.

Fob Watch

The fob watch, or pocket watch, was used by a number of the Doctor's incarnations. But it's the Tenth Doctor's fob watch that is of interest here. A priceless-looking antique, the Tenth Doctor's fob watch was able to return him back to his original personality when opened. If you recall, the watch kept his memories and physiology after he changed his appearance and personality to hide from the Family of Blood. The fob watch only worked for the owner, which meant anyone else opening it got only glimpses of the memories within. A Gallifreyan gadget to consign your USB flash drive to the trash can of history, and guaranteed to become a future broken promise of sci-fi. Incidentally, another imaginative creation from the Tenth Doctor story "Human Nature"/"Family of Blood" was the so-called Journal of Impossible Things. This was a dream diary that held the notes and sketches by the Tenth Doctor's human persona John Smith. The sketches seen on screen include those of the First, Fifth, Sixth, Seventh, and Eighth Doctors, the first time each had been depicted in the revived series. The journal also features sketches of the outside and inside of the TARDIS, a Sonic Screwdriver, K9, Cybermen and, naturally, Daleks.

Vortex Manipulator

Perhaps this invention takes all sci-fi inventions to their ridiculous limit. Get this: the wrist-mounted Vortex Manipulator, with the aid of the Sonic Screwdriver, could enable the wearer to travel up to one hundred trillion years through time. Sure. Decked out like a tiny games console strapped to your wrist, the Manipulator's main feature is its precision—it can hit an exact time, date, and location, apparently. What's the difference between the TARDIS and the Manipulator? The Tenth Doctor has an answer: the TARDIS is a "sports car," while the manipulator was more of a "space hopper." Don't go looking for these on Amazon or eBay.

Gadgets of Sci-Fi into Fact

Having looked at some of *Doctor Who*'s greatest gadgets, here are some inventions of sci-fi that made the real market.

1962: *The Jetsons* cartoon portrays phone calls accompanied by an image on TV. Since 2018, the Facebook Portal uses AI to provide a smart home device made for video calling. The Portal uses AI to pan and zoom, making sure the user is always on-screen.

1966: *Star Trek* inspired many tech companies to innovate now commonplace gadgets, such as Bluetooth headsets, cell phones, and automatic sliding doors.

1979: *The Hitchhiker's Guide to the Galaxy* "invented" the Babel Fish which, when placed in your ear, could translate any language by eating the language spoken to you and spitting out a translation of your own language into your ear. Since 2019, Google Assistant Interpreter Mode translates conversations in real time, and includes a lexicon of twenty-seven different languages.

1984: *The Terminator* featured self-aware AI weapons that attempt to wipe out humans with a nuclear strike. Since 2016, Google Translate uses AI to translate language pairs without explicit training, and ATLAS, an AI in development by the US Army, will identify and target threats. ATLAS does it all except pull the trigger. So far.

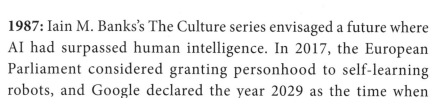

1987: Iain M. Banks's The Culture series envisaged a future where AI had surpassed human intelligence. In 2017, the European Parliament considered granting personhood to self-learning robots, and Google declared the year 2029 as the time when computers will surpass human-level intelligence.

1989: Marty McFly in *Back to the Future II* sees kids wear smart glasses that let them watch TV and answer calls, while Doc Brown wears a smart watch that can predict the weather down to the second. Since 2012, Dark Sky provides minute-by-minute weather forecasts, and in came the 2015 Apple Watch.

2002: Finally, the movie *Minority Report* showed an AI using personal data to target advertising to individuals. Since 2013, Tesco Petrol uses facial scans to determine a driver's age and gender, so they can target displayed ads as the customer fills their tanks.

What Are the Peaks of the *Doctor Who* Soundscape?

"French audiences watching *Doctor Who* dubbed into their native tongue obviously can't be expected to know the difference between Christopher Eccleston's Lancashire vowels and David Tennant's adopted southern accent—but isn't there some metaphysical conundrum posed by having the same voice actor, David Manet, play them both? Perhaps realizing this, Manet regenerated when Tennant did, and was replaced by Marc Weiss, who now brings mellifluous Francophonie to Matt Smith's performance. Le fez c'est cool."

—Tim Martin et al., "The Doctor lives longer in France,"
The Daily Telegraph (2015)

"I got my bunch of keys out, I got my mum's front door key, and scraped that up the strings (on the back of a piano). We did that several times on the bass strings on an old Sunday school piano that had been taken apart. So we took those and speeded them up, slowed them down, and cut several of them together, and started to add feedback to get that echoey sort of thing."

—Brian Hodgson, *The Daily Telegraph* (2013)

The Soundscape of Sci-Fi

In May 2017, the *Star Wars* website published an article about the revolutionary ways in which the 1977 movie *A New Hope* had "changed everything." Most of these ways were predictable enough; the speed of the *Star Wars* vehicles on screen, the new camera tech, and the lived-in look of the *Star Wars* Universe. But one of these revolutionary ways was a bit more surprising. It centered around the soundscape of *Star Wars*. George Lucas had gone on record about this topic before. He said that he felt around half of the unique cinema experience of *Star Wars* was down to the revolutionary use of both organic and electronic sounds. Sounds that had built the kind of soundscape moviegoers had never heard before. Chewbacca's growls came from bears, walruses, and other animals; the blaster sound came from slamming radio-tower cables and mixing the reverberation with bazookas; and the TIE fighter scream was made from an electronically stretched elephant howl.

Doctor Who was doing the same thing in 1963. From the very start the weird sounds began. Sounds which thrilled and terrified generations of viewers in equal measure. The sounds of alien planets, the terrifying voltaic-ranting voice of the Daleks, and the roar of the TARDIS. Steven Moffat was well aware of this sonic pedigree. In his commentary on the Twelfth Doctor episode "Listen," Moffat said the "Listen" story was *based* on sound. So here's an opportunity to sample the sonic delights from the *Doctor Who* archives. Once more, your journey through the history of the Doctor is well-suited to some supplementary surfing on YouTube.

TARDIS

To some, it sounds like a kind of cosmic elephant. To the latest *Doctor Who* production team, the noise of this police box-shaped spacecraft materializing from thin air is "VWORP" (at least according to the subtitles on the BBC during the first season of the Thirteenth Doctor). It's had its ups and downs over the years, but the TARDIS sound effect first appeared as part of the First Doctor story "An Unearthly Child" in 1963. As part of the BBC's Radiophonic Workshop, sound engineer Brian Hodgson made many of the iconic sound effects in the 1960s that have become firm favorites, including the TARDIS.

It seems that, despite being an inter-dimensional time-traveling machine, the noise of the TARDIS as it rips through the fabric of space-time has rather more mundane origins—a broken piano and a set of house keys. Hodgson admitted that at first he found it hard to imagine what a time machine would sound like. "It was quite difficult as everybody knew rockets went 'bang, whoosh' —but what does a time machine do? It doesn't go up, it doesn't go down, it goes everywhere at once. The thing I had in my mind was that it should be coming and going, and very vague." In the end, Hodgson tried to re-create for the TARDIS a noise similar to one he had made for a program called *The Survivors*, to represent a ship scraping on rocks. "So I put the rising note in it with loads of feedback and the TARDIS was born."

Doctor Who Theme

Composed by Australian Ron Grainer and also making its first appearance in 1963, the *Doctor Who* theme has been with us ever since, even though it's gone through many different arrangements and remixes. The original recording of the theme is widely regarded as a significant and innovative piece of electronic music, recorded way before the days of commercial synthesizers. Delia Derbyshire of the BBC Radiophonic Workshop realized the score written by Grainer.

Each note was separately created by cutting, splicing, and altering the speeds of analogue tape segments, which held record-ings of a single plucked string, white noise, and simple harmonic waveforms. The main, pulsing bassline rhythm was made from a recording of a single plucked string, played over and over again in different patterns. The swooping melody and lower bassline part were made by manually tweaking the pitch of oscillator banks to a carefully timed pattern. The non-swooping parts of the melody were made by playing a keyboard attached to the oscillator banks. And the rhythmic hissing sounds, "bubbles" and "clouds," were made by cutting tape sections of filtered white noise. Each note was then trimmed to length by cutting the tape, and sticking the sections together in the correct order. This was done for each "instrument" in the music—the two basses, the hisses, the swoops, the melodies, and the bubbles and clouds. Almost all of these separate bits of analogue tape making up the music still survive.

The story goes that Grainer was amazed at the finished *Doctor Who* theme when he first heard it. He asked, "Did I write that?" Derbyshire modestly replied, "Most of it." However, the BBC, who wanted to keep Radiophonic Workshop members anonymous, stopped Derbyshire from getting a co-composer credit and a share of the royalties.

The Daleks

BBC Radiophonic Workshop sound engineer Brian Hodgson also worked on the Dalek sounds. Hodgson has told how he was aiming for a "rather posh robot" voice. He then worked with Peter Hawkins, the voice of the Daleks and Cybermen in *Doctor Who* episodes in the 1960s. As Hodgson describes it, "Peter came up with the typical Dalek staccato and I asked him to elongate all the vowels because you only hear the modulation on the vowels. If you modulated the consonants they disappear and you can't hear it—so he did it on all the vowels and that became the Dalek voice."

There are other superbly innovated sounds associated with the Daleks. Other Dalek soundscapes the show has created in its history. For example, there's the sound of the "Dalek control room." And there's the sound of the Dalek City itself, a kind of metallic sound that gives you the impression of endless corridors on an alien planet. In fact, the Dalek city sounds are an example of the way in which budget constraints don't need to damage the science portrayed. What many viewers of the very first Dalek story recall was the striking sound of the Dalek city. It had a strange metallic resonance, overlaid with random sounds, which could have been the metal structure flexing as parts of it expanded and contracted, or the sound of some distant activity or machinery distorting into a weird sinusoidal waveform as it traveled down the metal corridors. Viewers were left with the impression of a living, and very alien, metallic entity.

The sounds were as much a signature for the Daleks as the way that the interior sounds of the TARDIS were a signature for the Doctor. This became more so when viewers heard the background sounds of the Dalek scanners and instruments in the Dalek control room, which followed the Daleks through many of their adventures.

The War Games

And that brings us to anther iconic *Doctor Who* sound—the "alien" sound in the Second Doctor story "The War Games." The story features the sounds of Gallifrey, home of the Time Lords. Mostly due to budgetary constraints, the BBC Radiophonic Workshop use a running sound loop of an eerie, pulsing drone (it's not easy to describe, so this is where some supplementary surfing on YouTube is useful!). The final result is suggestive of the vast and empty spaces on the alien planet beyond the dark walls.

Listen!

In the Twelfth Doctor episode "Listen," the Doctor and Clara are about to take Orson Pink off the space station. But what is that creepy noise they hear? What's about to attack them? Is it an alien, or maybe some kind of monster from deep space? Or perhaps simply the sounds of a space station going through its motions in the dead of night. "Listen" has been called the most conceptual episode in the entire history of *Doctor Who*. The story is essentially about a monster we never see and might not even exist. The Doctor conceived of the monster's existence in a weird way. Considering that nature had evolved creatures who are primed for hunting and defending themselves, the Doctor theorizes that evolution would

also produce an animal with peerless camouflage. Consequently, this apex camouflaged creature would leave no evidence of its own existence. There are a lot of creepy noises in this story, especially on Orson Pink's space station at the end of time. And what might make creepy noises? Perfectly camouflaged creatures, that's what.

Part IV
Monster

Introduction

"A lot of our heroes depress me. But you know, when they made this particular hero, they didn't give him a gun, they gave him a screwdriver to fix things. They didn't give him a tank or a warship or an X-wing fighter, they gave him a call box from which you can call for help. And they didn't give him a superpower or pointy ears or a heat ray . . . they gave him two hearts. And that's an extraordinary thing. There will never come a time when we don't need a hero like the Doctor."

—Steven Moffat, The Doctor Who 50th Celebration Weekend
(November 24, 2013)

What can we learn from the monsters of the Whoniverse? As with the three parts before this, our organizing idea for this final part shall be that sci-fi monster stories put humans up against the non-human. But, as always, appearances can be deceiving. Our idea of what is normal gets morphed out of all recognition when it comes to monsters. We might think of monsters as bug-eyed beasts. But on closer inspection, the boundaries begin to blur. Take the Doctor, for example. As Steven Moffat says, the Doctor is pretty much a superhero, of sorts. Coming to the rescue with two hearts beating, and the ability to literally cheat death. Sure, he *looks* human. But it's merely a mask that hides a darker, more alien nature within. It reminds us that behind our civilized exterior lurks a monster, waiting to run amok. And often, to meet the real monsters in sci-fi, humans merely have to look in the mirror.

Since the days of Darwin, the human future has fascinated sci-fi. Writers and moviemakers have wondered what might become of us humans. And what might become of life itself in this new Universe. Over the last 160 years since the publication of *Origin of Species*, thought-provoking sci-fi has focused on two aspects of biology: evolution and genetics.

No such stories were written in the Age of Faith. Everything from architecture, art, literature, and music were based on the confidence that God made the cosmos for us. But, after Darwin, we found ourselves among the microbes with no special immunity from natural law and with vanishingly little evidence of a divine image. American astronomer Carl Sagan described the progress of science in the last five centuries as a series of great demotions. We humans have found ourselves on a mere planet, in orbit around an ordinary star, adrift in a galaxy among a trillion more. Each successive discovery seemed to impact not just the human condition, but also the meaning of life in the Universe. And with techno-science now unraveling the human genome, the twenty-first century may hold even greater change. We face a future of conscious intervention in reproduction and heredity and directed evolution. Little surprise that this brave new world is sometimes dubbed "the Frankenstein century." Perfect hunting ground for the tales of the Doctor.

Doctor Who has presented, on the one hand, compelling projections of our future based on the theory of evolution. While, on the other hand, it has also told mesmerizing tales about the remaking of man through the mind-boggling power of genetics. *Doctor Who* has provided a sustained, coherent, and often subversive check on the contradictions of science, the promises and pitfalls of progress through the ages. Those stories have reflected advances in biology. But *Doctor Who* has calculated the human cost of the darker aspects of those advances in natural science.

So *Doctor Who* monster sci-fi is mostly about that definitive terror: ourselves. The monster that goggles back at us when we stare into the mirror. Homo sapiens, predator par excellence. A creature so deadly that what they do to each other is far worse than what can be done to them. Monster stories in *Doctor Who* are often an allegory of our concerns for humanity's fate, in a world created by us.

Sometimes *Doctor Who* turns to classic monsters, and gives them that "sci-fi" twist. Weird wolf-men become the victims of a disease called lycanthropy, a warning perhaps of technological tinkering with biochemistry. In the case of the werewolf in the Tenth Doctor story "Tooth and Claw," it is an alien werewolf, used by a group of warrior monks who intend to take over the British Empire. *Doctor Who* seldom disappoints in terms of ambition. Zombies become the result of viruses escaped from the laboratory of a mad scientist, or perhaps a wry comment on submissive humans in a society dominated by the culture of consumerism. In the case of the Ninth Doctor tale "The Unquiet Dead," the Doctor teams up with none other than Charles Dickens to investigate zombies associated with a funeral home where the gaseous Gelth is reanimating corpses through a space-time rift. Cybermen are also a form of the undead, particularly in stories like "Tomb of The Cybermen."

According to *Doctor Who*, we can't hide it forever, this gothic at the heart of the human. It lurks beneath the mask, a creature ready to erupt, the Jekyll within the Hyde. So many stories remind us that this culture of science and civilization that we created is a fragile masquerade. Like the DNA we've begun to tamper with, our cunning construction of society could unravel at a moment's notice. *Doctor Who* deploys Daleks, Cybermen, and clones to tear down this precarious contrivance.

Yet for every action, there is an equal and opposite reaction, even if it comes clad in a police box. The superhero solution is ever ready. The Doctor variety doesn't have rippling muscles. The Doctor isn't armed with laser vision, superstrength, or the power of flight. But the Doctor *does* have gallons of righteous indignation. The benevolent monsters of the Doctor come not from the id but the ego. The Doctor stands as a totem to our determined ability to overcome obstacles and renegotiate peril. Forever willing to deliver an impromptu lecture on good winning out over evil, the Doctor always conspires to enable the monster to defeat itself.

So turn over the page and read this Monster section, where the relentless march of the monster continues. Though you may meet a carnival of monsters to terrify and traumatize, with bug eyes, huge pincers, and hairy bodies, remember that monstrosity is just as much a feature of the mind as it is of the body. It's time to come out from behind the couch and confront Sontaran girlfriends, a history-stalking Silent, and the prospect of defecating Daleks.

Is He a "Mad" Doctor?

"Internal memos from the BBC's original series, unearthed in 2010, describe the 'metaphysical change' of the Doctor's regeneration as 'an experience in which he relives some of the most unendurable moments of his long life, including the galactic war.' (No doubt the writers of the John Hurt arc leapt in joy to read that.) 'It is,' continue the memos, 'as if he had had the LSD drug and instead of experiencing the kicks, he has the hell and dank horror which can be its effect.'"

—Tim Martin et al., "The Doctor's regeneration was based on a bad acid trip," *The Daily Telegraph* (2015)

Cinema Scientists

Scientists have had a tough time of it on the silver screen, television as well as movies. And the gap between the expert knowledge of the scientist and the public understanding of science serves at the source of most fictional representations of the scientist. That gap had usually been filled by clichéd and stereotypical characterizations of one type or another. Or, as one MGM executive wrote to Albert Einstein in 1946 about a screenplay for a highly fictionalized tale of the first atom bomb, "it must be realized that dramatic truth is just as compelling a requirement on us as veritable truth is on a scientist." One assumes such "dramatic truth" was the way in which the expert knowledge of the scientist was made to "come alive" to mass audiences.

In time, we began to realize just how potent the dramatic truth of those "Hollywood" images of the scientist could be. Six years before *Doctor Who* began, American anthropologist Margaret Mead carried out a pioneering survey of thirty-five thousand American high school kids. She was interested in how school kids tend to draw or describe their mental image of the scientist. The kids were asked to complete the sentence "when I think about a scientist, I think of . . . " The results from the 1957 study came out like this:

> The scientist is a man who wears a white coat and works in a lab . . . he may be stooped and tired. He is surrounded by equipment: test tubes, Bunsen burners, flasks, and bottles . . . he experiments with plants and animals, cutting them apart, injecting serum into animals . . . he is a very intelligent man—a genius or almost a genius . . . if he works for the government, he has to keep dangerous secrets . . . if he works for a big company, he has to do as he is told; he is just a cog in a machine . . . he may even sell secrets to the enemy. His work may be dangerous. Chemicals may explode. He may be hurt by radiation, or may die . . . he bores his wife, his children, and their friends with incessant talk that no one can understand . . . he is never home.

The results of further studies over the next forty years became worryingly consistent. Mead's initial conclusions had proved difficult reading. These images of the scientist held very little attraction for young Americans. The main culprit? The mass media image of the scientist. Their portrayals were of either "cogs in the machine," or "isolated" loners. But, despite publicity of the surveys, the same stereotypes remained for decades. Sure, they evolved, but they also abided. American cultural commentator Theodore Roszak put it like this:

> Asked to nominate a worthy successor to Victor Frankenstein's macabre brainchild, what should we choose from our contemporary inventory of terrors? The Bomb? The cyborg? The genetically synthesized android? The despot computer?

Modern science provides us with a surfeit of monsters, does it not? I realize there are many scientists—perhaps the majority of them—who believe that these and a thousand other perversions of their genius have been laid unjustly at their doorstep. These monsters, they would insist, are the bastards of technology; sins of applied, not pure science. Perhaps it comforts their conscience somewhat to invoke this much-muddled division of labor . . . Dr. Faustus, Dr. Frankenstein, Dr. Moreau, Dr. Jekyll, Dr. Cyclops, Dr. Caligari, Dr. Strangelove. The scientist who does not face up to the warning in this persistent folklore of mad doctors is himself the worst enemy of science.

Or as American cultural historian David Skal put it, "The mad scientist is so ingrained in modern thought, that you almost don't notice he is there." Is the Doctor to be added to the list from Frankenstein to Strangelove? To what extent has our Doctor been part of this persistent folklore of mad doctors? The gap between the expert knowledge of the scientist and the public understanding of science goes back to the scientific revolution. That's when ideas like "technical" and "popular" first came to the fore. That gap has since been filled by saints like Isaac Newton and sinners like Victor Frankenstein. But *Doctor Who* began merely a few years after Mead's initial study. Has the Doctor been guilty of the same persistent folklore of mad doctors?

"Mad" Doctor?

According to academic Roslynn Haynes, a series of sequential images of scientists have become deeply embedded in our culture. Those images of cinematic stereotypes are:

- *The Alchemist* (late sixteenth century): the scientist who seeks arcane and forbidden knowledge, mostly works alone, and is driven by a madness for power, tending to be intellectually arrogant.

- *The Absent-Minded Professor* (mid-seventeenth century): an obsessive scientist, a reductionist who narrows science down to the single-minded search for a tiny aspect of knowledge, so much so that he neglects social responsibilities.
- *The Inhuman Rationalist* (early nineteenth century): a suppressor of human emotions in favor of detached enquiry, and who ignores the social dimensions of scientific findings.
- *The Heroic Adventurer* (late nineteenth century): intrepid researcher who boldly goes where no one has gone before, and who may be rather eccentric.
- *The Helpless Scientist* (mid-twentieth century): well-intentioned researcher whose findings are hijacked by government or corporate concerns.
- *The Social Idealist* (mid-twentieth century): influenced by late H. G. Wells, this idealist is driven by social conscience rather than hard research. Their maverick heroism stems from non-compliance with government or corporate concerns.

Until the late twentieth century, writers have overwhelmingly condemned science and scientists. And many stories of scientists stuck to one or more of the stereotypes above. But, since the 1990s, science has acquired a more humane image. Scientists are now rarely objects of fear or mockery. Scientists combat deadly viruses. Scientists save endangered species. And scientists communicate the climate crisis. Novelists have become more intent on authenticity in presenting science and the motives and moral dilemmas of scientists.

And, we might argue, *Doctor Who* was ahead of the curve. From the start, the Doctor was portrayed more in the kindly vein of the heroic adventurer or the social idealist. Take a look at a selected list of Doctors below and see if you agree with my characterizations.

First Doctor: William Hartnell

Frail-looking and slightly aloof, the First Doctor assumed many of the characteristics of the Victorian Gentleman Scientist. At first he was mistrustful and treacherous, and almost seems to have been created as a villain in the new series, especially after he kidnapped companions Ian and Barbara. But what we sometimes forget is that this Doctor was on the run. He was a political refugee in hiding and he couldn't afford to leave potential witnesses behind for the Time Lords to find. In fact, he'd been on the run so long, while also looking after his granddaughter Susan, his nerves must have been near breaking point. It's only after his companions prove to be reliable friends and allies that his attitudes began to soften. Within his new extended family he began to enjoy exploring the Whoniverse much more.

Verdict: Heroic Adventurer. He exhibited a sense of place and dignity. When a policeman asked if he was British, the First Doctor replied, "I am a citizen of the Universe, and a gentleman to boot!"

Second Doctor: Patrick Troughton

By all appearances, the Second Doctor seemed clownish, but under the surface lurked a smart and far cannier character. After many adventures, the Time Lords condemned him to Earth for messing in other planets' affairs. He had a strong sense of justice. If something was wrong then it was wrong and should be interfered with. He put up a persona of confusion and incompetence that was hard not to believe. His being prone to thinking out loud only added to this impression as he questioned his own options. Kind and caring, it's difficult not to like this Doctor.

Verdict: Helpless Scientist/Social Idealist. An excellent trickster, tricking humans, Daleks, and other enemies into doing things his way.

Third Doctor: Jon Pertwee

The Third Doctor was exiled by the Time Lords, so he spent his time stuck on Earth. It's worth noting here the influences of TV Production and budgets. The TARDIS was grounded by the BBC to save money on expensive alien sets and locations, so something was needed to replace the specialness of the TARDIS, which was Bessie and the Whomobile.

Verdict: Heroic Adventurer/Social Idealist. A kind of dashing gentleman adventurer with an air of superiority, a man of action, a kind of alien James Bond who had his own archnemesis in the form of the Master—the Third Doctor had a heart of gold and was very patient and nurturing of his female companions.

Fourth Doctor: Tom Baker

For many, the most recognizable Doctor, kitted out with iconic hat and scarf. After an initial story involving UNIT he turned his back on a job working for the authorities, preferring to wander through time and space always eager to look for trouble. The Whoniverse didn't feel so parochial with the Fourth Doctor. His time included some of the best stories and adventures. Filled with a license to roam after being stuck on Earth, his watch included not only Sarah Jane Smith and Leela, but also the first appearance of the robot dog, K9. A first-class telepath, he was able to read the minds of others without contact.

Verdict: Heroic Adventurer/Social Idealist. He had a slightly mad persona as he confused potential enemies with his unconventional behavior, like taking a hostage and threatening him with a jelly baby (candy). Once this would-be captive agreed to his terms, the Doctor then casually ate the candy in front of him. The Doctor refused even to carry a weapon (hence the use of a jelly baby) much less use it. Moved to anger at unnecessary killing, he was even reluctant to wipe out the nascent Dalek race. Despite his unusual personality, he was a very logical Doctor.

Ninth Doctor: Christopher Eccleston

The Ninth Doctor was a survivor of The Last Great Time War, an event that haunted him during much of this incarnation. This was the Doctor who met companions such as Rose Tyler and Jack Harkness, who seemed to help lift him out of the brooding darkness.

Verdict: Social Idealist. He was a Northerner ("Lots of planets have a north") who dressed a little like a U-Boat Commander, and was something of a working class scientist, in the same way Superman is a friendly neighborhood superhero. One can imagine the Ninth Doctor throwing a Molotov cocktail at an oppressive regime or invader. He never seemed comfortable being a character that people looked to for guidance or leadership. Perhaps he had seen those qualities abused or misused too often?

Tenth Doctor: David Tennant

David Tennant played the most charismatic of Doctors. The Tenth Doctor had a great appetite for life, and a strong sense of being the last of the Time Lords. An excellent scholar, he was able to rewrite his biology, storing his Time Lord characteristics in a watch, so he could turn human. His Alienese was highlighted by this, as was his consideration for the aliens hunting him. But when they killed innocent people, his vengeance was terrible and considered. He gave the aliens what they wished for, but in a way that gave meaning to the old expression "Be careful what you wish for!"

Verdict: Social Idealist. Very human in his relationships with his companions Rose, Martha, and Donna, he met his end after absorbing a vast amount of radiation when saving Donna's granddad.

Eleventh Doctor: Matt Smith

The Eleventh Doctor was a most charming Doctor, and one with a notable number of achievements. He is the Doctor's most long-lived incarnation, the final incarnation of the first regeneration cycle, and one who had a centuries-long struggle against his enemies, including the last stand on Trenzalore.

Verdict: Heroic Adventurer/Social Idealist. Looking young enough to make you believe he should still be in school, the Eleventh Doctor certainly broke the crusty old stereotype of an aged academic. He exhibited a youthful exuberance and sense of wonder at the Universe around him. He sometimes seemed to rush headlong into situations that left him scrabbling around for a solution. But he always seemed to get that inspirational idea when things looked their bleakest.

War Doctor: John Hurt

The War Doctor, also known as the Renegade, was the warrior incarnation of the Doctor who lived a very different timeline compared to his other lives. As a fighter, rather than a peacemaker, he managed to bring an end to The Last Great Time War.

Verdict: Heroic Adventurer. A fighting and battling Doctor who, unlike his other incarnations, seemed to embrace the use of weapons as a means to an end. He was bred (so to speak) for war, but he is still the odd one out among the Doctor's incarnations. He did what none of them could have done, wiping out the Daleks and the Time Lords in order to end a destructive war that was damaging all of space-time.

Twelfth Doctor: Peter Capaldi

After battling in the Siege of Trenzalore for nine hundred years, the Doctor was facing extermination, but Clara appealed to the Time Lords, and the Doctor was gifted a new regenerative cycle.

The Twelfth Doctor that emerged is a darker character, who often appeared fearsome and ruthless.

Verdict: Heroic Adventurer. The Twelfth Doctor behaved like someone enjoying a new lease on life, willing to try new things and appreciating all living things in a way he never did before. He treated the animals he spoke to as if they were his equal, sometimes seeming to prefer them to the humans he encountered. As well as his immediate companions, he had an extended group of allies known as the Paternoster Gang, made up of Madame Vastra, Jenny Flint, and Sontaran Strax. Sometimes prone to impatience and sarcasm and not gladly suffering fools, he could be brutally judgmental and unpredictable.

Questions for the Doctor:
Will Humans Evolve into Daleks?

In the Fourth Doctor story "Genesis of the Daleks" (1975), the Doctor intervenes in the creation of the Daleks to prevent a future in which the Daleks rule the Universe. Davros tells the Doctor that he [the Doctor] has a weakness that has been totally eliminated from the minds of the Daleks so they will always be superior. That weakness is a conscience.

> "As Ruskin has said somewhere, a propos of Darwin, it is not what man
> has been, but what he will be, that should interest us."
> —H. G. Wells, *The Man of the Year Million* (1893)

Daleks and Martians

Will humans one day evolve into something like the Daleks? The idea of the Daleks was very much influenced by the portrayal of the Martians in H. G. Wells's 1898 book *The War of the Worlds*. Like the Daleks, the Martians in Wells's book are frail in body, but mentally very intense. The Martian machines are operated by the crab-like creatures that designed them, their weak tentacles moving feebly over steampunk controls. It sounds very similar to the frail Kaleds inside those rather steampunk Dalek shells.

Again like the Daleks, the Martians have superior technology with which they are able to dominate humans. They use sophisticated tech in their vast journeys across space. The Martians have heat ray weapons and poison gas, and they suggest that all attempts at resistance are futile. Like the Daleks, the Martians are an unrelenting force of nature, and agents of the void. They make it clear that, in the cosmic chain of command, *they* are in charge.

But the immense coldness and indifference of these aliens, Daleks and Martians alike, are also a glimpse of what humans may one day become. Like us they struggle along in a hostile Universe, a cosmos vast and cool and unsympathetic. They are also a glimpse of our future in terms of their over-developed brains and under-developed bodies. They are the tyranny of intellect alone.

The Man of the Year Million

So, the Daleks are a vision of a future humanity. The first of Wells's great novels, *The Time Machine*, also plotted a dark future for humanity, and had a huge influence on both Daleks and TARDIS. *The Time Machine* has two major themes: evolution and social class. The book is an ingenious voyage of discovery through the invention of a machine, which symbolizes the power of science and reason. The Time Traveler sets out to navigate and dominate time. His discovery? Time is lord of all. The significance of the story's title becomes clear. Humans are trapped by the mechanism of time, and bound by a history that leads to our inevitable extinction.

The Traveler's headlong fall into the future begins at home. The entire voyage through the evolved worlds of man shows little spatial shift. The terror of each age unravels in the vicinity of the Traveler's laboratory. "It is not what man has been, but what he will be, that should interest us," Wells had written in his 1893 essay, *The Man of the Year Million*. In *The Time Machine*, and in the form of the Daleks in *Doctor Who*, we have our answer—a vision calculated to "run counter to the placid assumption . . . that evolution was a pro-human force making things better and better for mankind."

In *The Time Machine*, time's arrow thrusts the story forward to the year 802701 AD. The Traveler meets the Eloi, a race of effete, androgynous and childlike humans, living an apparently pastoral life. Humanity's conquest of nature, it seems, has led to decadence.

On discovering the subterranean machine world of the albino, ape-like Morlocks, a new theory emerges. Over time, the gulf between the classes has produced separate species.

It is the technological Morlocks that provide a model for the Daleks. Whereas with the Eloi "upper-world man had drifted towards his feeble prettiness," the Morlock future was an "under-world [of] mere mechanical industry" whose "perfected science… had not been simply a triumph over nature, but a triumph over nature and the fellow-man." This comparison of Morlocks and Daleks makes even more sense when you consider the short story that Dalek-creator Terry Nation wrote in 1973. Entitled *We Are The Daleks*, the tale is the first time Nation gave an origin story for the Daleks. Nation revealed that the Daleks were humans from some point in the future. They were the result of an accelerated and engineered evolution.

Nation's idea for the Daleks can be seen in Wells's *The Man of the Year Million*. Wells claimed that, in terms of evolution, there is little difference between Stone Age humans and modern day humans. The only real difference, though huge, is the artificial divergence based on the available science and tech. If transported to a modern metropolis, Stone Age humans would have little chance of understanding modern civilization. Might the same also be true of modern humans? The Man of the Year Million will supersede modern humans based on a similar divergence through the artificial evolution of science and tech, placing the usual laws of evolution to one side.

In short, tech trumps nature, Daleks trump humans. After 1893, having written the essay on an artificially-evolved future, Wells then began to imagine alien life-forms that might have made such a decision in the distant past, and how they might look now. He considered the fate of the Tasmanian people. In the late 1800s, the Tasmanians had loomed large in the conscience of the civilized world. They were the first race deliberately driven to extinction by colonists who met their ancient spears with modern guns. Wells imagined how an alien race of similar technological superiority might appear to us.

Wells imagined Martians, *Doctor Who*'s Terry Nation imagined Daleks. And both these acts of the imagination stem from Wells's journey that began when he envisaged the final stage of humanity in *Man of the Year Million*. Giant brains, and puny bodies reduced to vestigial sacs, except for super-sensitive hands to transmit the brain's instructions to the advanced tech that had become their exoskeleton. It's the Daleks, no doubt.

Puny Humans

And indeed, according to a report by *National Geographic* in 2014, we humans already appear to have evolved puny muscles. We did this even faster than we grew big brains. This is according to a metabolic study that pitted people against monkeys and chimps in contests of strength. Like the Daleks, weak muscles is the price we have paid for the metabolic demands of our amazing cognitive powers. Our early ancestors also possessed apelike strength. At least for the skeletal muscles looked at in the *National Geographic* study. Today human brawn is much reduced. Meanwhile, other body tissues, such as kidneys, have remained relatively unchanged over millions of years.

So Wells was right. Geeks rule. And geeks may become Daleks. The major difference between modern humans and other apes, such as chimps, is our oversized, energy-hungry brain. It was the evolution of the human brain that drove the adaptation of our early ancestors away from an apelike ancestor, starting roughly six million years ago. The study seems to confirm the idea that weak muscles, along with a weakness for the couch—conducive to brain-work like reading and watching *Doctor Who*—is our evolutionary inheritance. If this trend continues into the distant future, in what ways will our future form be made weird and wonderful by technology? The Dalek invasion of Earth has already begun.

Is the Doctor a Superhero?

"I teach you the overman. Man is something that shall be overcome. What have you done to overcome him? . . . All beings so far have created something beyond themselves; and do you want to be the ebb of this great flood, and even go back to the beasts rather than overcome man? What is ape to man? A laughingstock or painful embarrassment. And man shall be that to overman: a laughingstock or painful embarrassment. You have made your way from worm to man, and much in you is still worm. Once you were apes, and even now, too, man is more ape than any ape . . . the overman is the meaning of the earth. Let your will say: the overman shall be the meaning of the earth . . . man is a rope, tied between beast and overman—a rope over an abyss . . . what is great in man is that he is a bridge and not an end."

—Friedrich Nietzsche, *Thus Spoke Zarathustra* (1883)

Superheroes

Superheroes appear in most cultures. Japan has Ultraman. Mexico has Santo. Pakistan has Burka Avenger, and South Africa has Jet Jungle. And the Doctor belongs to a rather long pedigree of "superheroes" in British culture. Shakespeare's sceptered isles have created many such heroes over the years: King Arthur, the most mythical of all monarchs; Robin Hood, the most legendary of all outlaws; Sherlock Holmes, the definitive consulting detective; James Bond, the archetypal super-spy; and Harry Potter, the world's most powerful young wizard. And, naturally, the Doctor—hero of the world's longest-running sci-fi series. But is the Doctor really a superhero?

We could compare the Doctor to that other famous alien superhero—Superman. One of the most enduring icons in the American cultural landscape, Superman first made an appearance in *Action Comics* in 1938. And, like the Doctor, this powerful fictional figure of the twentieth century has often adapted to changing times. When World War II broke out, Superman's slogan changed from fighting for "truth and justice" to fighting for "truth, justice and the American way." This new motto lasted into the 1950s, when Superman became a symbol of a robust American chauvinism, which could do little wrong. And, as the darker moral relativism of the twenty-first century has crept into our culture, Superman has adapted to this brave new world, uncertain and fearful of the fall, whether economic, political, or environmental. In the 2013 movie *Man of Steel*, after saving a busload of schoolmates from drowning, a distressed teenage Clark Kent confides in his stepfather, who is worried Clark has given away his alien nature, asking "What was I supposed to do? Just let them die?" His stepfather's reply? "Maybe."

Sometimes the Doctor too behaves in what seems like a totally immoral way. For Superman fans, it must have come as a shock if his popularity was solely down to the result of his embodiment of goodness, of what is right and wrong. In this sense, both the Doctor and Superman as people fulfill the same societal function as the myths of ancient Greece or Rome. Some folk need myths to teach them virtues. And those virtues may need to be embodied by a person. But if that person truly captures the notion of the Platonic ideal of the good, fighting the fight for high ideals and teaching moral lessons, where on earth is the world going if even the Doctor or Superman sacrifices human lives to save their own skin? For some, the Doctor and Superman are not merely superheroes flying high above us. They are also seen as supernatural forces, transformed into Christ-like figures, who descend to Earth from the skies, bringing heavenly powers as they walk among us mere humans.

Superhero Origins

Earlier in the book, we talked about the way in which the days of Darwin led sci-fi straight to the idea of the modern alien. In a similar way, theories of evolution have also given writers a creative framework for stories of super-humans. Sci-fi has a long and proud history, but it was only after Darwin that the hopes and fears for biology really developed. From ancient times, fantastic tales had always been told. But from the age of discovery to this Frankenstein century, those tales have reflected the new and ongoing relationship between potential progress in science and the skeptical echo of its fiction. In that way we can think of sci-fi as the literature of change since the Renaissance. It's given us a cultural commentary on the accelerating pace of change in society, and continues to use the fantastic to make sense of the dark "magic" of nature.

So, the question of our evolutionary future has busied the fertile brains of sci-fi writers and artists, and later filmmakers, ever since those early days of Darwin. In his own particular take on Darwin, German philosopher Friedrich Nietzsche floated the notion of the übermensch ("superman," "overman," or "superhuman") in his 1883 book *Thus Spoke Zarathustra*. Nietzsche's idea of the übermensch was of a being seeking to move "over" its state of being to a greater "stature." Nietzsche's recipe for the human future was a potent one. Few other symbols in sci-fi have evolved as dramatically as the "superman." From the most infantile form of human wish-fulfilment to the more sophisticated antihero, the superhero has become a playful metaphor of our aspirations and fears for future science. It is within this tradition, from Nietzsche to N'Jadaka, that we should consider the Doctor status as an übermensch.

Is the Doctor a Superhero?

Let's consider the evidence. We have already looked at the cultural pedigree of what it means to be a superhero. Not only that but, according to the good offices of Google, the Doctor without a doubt

qualifies for superhero status. Google tells us that a superhero is a character with extraordinary or superhuman powers and is dedicated to protecting the public. Google also says that superheroes may include any kind of fantasy fiction crime-fighting character.

Well, the Doctor is a humanoid-alien Time Lord from the planet Gallifrey, who can travel through time and space using an internally massive time machine. That's a good start in terms of superhero pedigree! He's also telepathic, has an ability to regenerate, has two hearts, a huge intellect, and often shows a superhero level of stamina. Like Superman, the Doctor has had his home planet destroyed, though the ultimate fate of Gallifrey is yet to be discovered. Like Batman and Iron Man, the Doctor is decked out in expensive tech—an essential part of their superhero status. And as there's a fine line between being a hero and a vigilante, like Spider-man and Batman, the Doctor may be just one adventure away from being put on trial for taking justice into his own hands. While other superheroes defend a relatively small patch, such as Gotham or planet Earth, the Doctor has the whole Whoniverse with which to contend.

Finally, consider the Tenth Doctor story "Family of Blood." Schoolboy Timothy Latimer gets an A++ for his plea to John Smith, the human version of the Doctor. When it's time for Smith to morph back into the Doctor to save the world, Timothy's compelling description hits the proverbial mark. Timothy's words paint a portrait of a classic troubled soul—a perfect storm of rage and wisdom, pain and passion. And yet it also reminds us that Timothy describes no ordinary human. The Doctor is near immortal and relatively omnipotent, with life experiences and a wisdom far beyond human comprehension. The Doctor is an übermensch.

How Do Daleks Defecate?

"Since making their appearance in the second-ever *Doctor Who* episode, the Daleks have become icons. And if you had had a spare £20,400 lying around in 2009, you could have got your hands on the only remaining 1963 Dalek (one of a set of four, together costing £1,000 to build), which was sold by auction house Bonhams as part of a large *Doctor Who* sale. A blue shirt worn by David Tennant went for £1,200."

—Tim Martin et al., "Daleks are collectors' items,"
The Daily Telegraph (2015)

"Over the years, the Oxford English Dictionary has made space between its august e-covers for several Whoisms. "Dalek" (now sometimes "used allusively," the OED informs us) was the first one; it's a word Terry Nation claimed to have made up on the spot, only later discovering that it meant "far-off" in Serbo-Croat . . . the series also gifted to the British language in perpetuo the phrase "hiding behind the sofa" as an expression of TV-induced fear. Whoever coined it, the formula is now inextricably linked to the Doctor (and the appearance of the Daleks), lending its name to a book of celebrity recollections of the series as well as a fallback headline for hacks everywhere. Even Prince Andrew has admitted to hiding from the Daleks behind the soft furnishings as a child in Windsor Castle."

—Tim Martin et al., "*Doctor Who* has changed the English language,"
The Daily Telegraph (2015)

Dalek Backstory

Planet Skaro, twelfth planet from its sun, is situated in the Seventh Galaxy and is home to the Daleks, survivors of a catastrophic nuclear war who have reengineered their species into cyborgs with protective robotized shells. The tank-like shells and Nazi xenophobia shown by the creatures inside are the brainchild of archetypal mad scientist Davros. Davros is boffin-in-chief of an elect cadre of scientists charged with the task by their despairing people, the Kaleds, to tip the tech balance and end their Thousand Year War with the neighboring Thals.

The first Dalek story was called simply "The Daleks." As the plot unfolds, we learn about their apocalyptic "neutronic war." We learn about their retreat into metal shells, their hiding place in which their emotions shrivel. This idea that they were once better "people" makes them more horrifying still, as we dread becoming like them. For viewers in 1963, Terry Nation's script would have been scarier than it seems today. The Cuban Missile Crisis was still fresh in viewers' minds. The surface of Skaro looked like the kind of world Earth might become after a nuclear war. And the "neutronic war" refers to the specter of the neutron bomb, a real human weapon designed to emit more radiation than an atomic bomb, but with a lower blast. The result? The neutron bomb wiped out human and animal life, but not buildings and infrastructure. In 1963, the Daleks seemed like a very real nightmare, the kind of creatures that could have designed the neutron bomb, highlighting once more the similarity between us and them.

Later we find out how bad Davros had been. He spends most of his time finding and weaponizing a way in which the Kaleds can live on beyond the end of the war. (Shades of the thousand year Reich.) The Daleks' backstory is essentially a military R&D enterprise. It's the same here on Earth. War has been a major trigger of tech, and in wartime money and people are thrown at tech in ways they simply aren't during peacetime. We can read the Kaled Technical Elect as a narrative based on the allied World War II project to build an atomic weapon before Hitler.

What about the science and tech of the Daleks themselves? Once they say goodbye to their humanoid Kaled origins, it's an entirely new ball game, being a mysterious creature, locked in the dark within a robotic shell. We learn that they become a space-faring, and even time-faring, species. We learn also that they conquer and enslave planets across the Whoniverse. And irrespective of the defeats meted out to them, the Daleks always seem to recover. Indeed, the Daleks become so potent a force that they bring the Time Lords to the jaws of defeat in The Last Great Time War. In the face of all these incredible, albeit fascistic advances, what kind of technological progress did they employ to ensure comfort within those protective robotized shells? In short, how did the Daleks, from the ordinary drone to the Emperor, defecate?

Dalek Anatomy

It is, of course, wildly apocryphal and mildly juvenile to suggest that one of the mitigating factors in the infamously bad temperament of the Daleks might be their inability to achieve biological release. But let's consider Dalek biology. To answer what is after all the scientific question as to how Daleks defecate, we first have to think about the very nature of the Dalek, and then think about how its biology might work.

What actually constitutes a Dalek? You've no doubt seen the mutant Dalek creature inside the robot shell. If not, Google "inside a Dalek" right now. You see the anatomy plainly. What most people think of as a "Dalek" is the machine that carries these creatures around, but the actual "Dalek" beings live inside the shell. One hundred percent engineered creatures of pure hatred for all other life forms. The machine shell is little more than a battle tank. It hides a war computer and a support system to keep alive the creature in that battle tank. It sometimes looks like the machine has only one weapon, but the entire entity can be used as a living bomb if necessary.

The Daleks themselves have no gender. They're neither male nor female. Deep down, hidden in their biology, Daleks may still have some way of giving birth to live young, just as humans do, as Daleks were originally created from human-like beings. In the labs where Daleks are made, they use cells from Daleks to clone other creature Daleks. So, one Dalek can produce countless thousands of copies of itself during its lifetime. As a single Dalek is said to live to be one million years old, it means that if only one Dalek survives, it's capable of booting up an entirely new Dalek race by itself. This is no doubt one of the "secrets" to the Daleks' survival and their ability to recover from seemingly irretrievable losses.

That familiar battle tank/robotic shell casing can be tweaked to make a support system. After all, the design is based on Davros's own life support mobility scooter. Since the Dalek is a "cyborg," this system has to join up the "org" bits of the Dalek with the "cyber" bits of the robotic shell. The Daleks are able to "have lunch" because the support system would include some sort of intravenous nutrient stream that is fed directly into their organic parts. In the same way, their waste products would be removed, like a dialysis machine works for people with kidney problems.

To make sure a Dalek doesn't defecate on duty, they could have control over the emptying of their waste tanks. Maybe, like trains parked at stations, Daleks are not allowed to defecate while stationary. Perhaps they defecate only when in motion. Your mission now, should you choose to accept it, is to look back over the *Doctor Who* catalog of episodes (currently standing at 851 episodes with 97 missing). Nominee researchers might like to pay close attention to see if they can spot a Dalek voiding its tank while on the go.

How Does *Doctor Who* Foresee a Future of Villainy?

In the Third Doctor story "The Time Monster" (1972), the story is set in a village near Cambridge as well as the mythical city of Atlantis. In the tale, the Master seeks the power of Kronos, a being that exists outside of time and space, so that he [the Master] can control the Universe.

> "Aren't we all interested in dominating other people? Total control, a machine-like desire, and the Daleks. It had a neatness that fitted. In the past only human endeavor has overcome this kind of evil. Man is a supreme creation if he is able to work with his fellows. There is nothing he can't do if he gets down to it. I believe that what people want on television is entertainment, and action stories are what I want to write."
> —Terry Nation, "*Doctor Who*'s Who Archive," *The Guardian* (1966)

Super-tyrants

Doctor Who stories are based on the ways in which science may transform our lives in the future. The ways in which we may be able to one day transcend our limits of body and brain. Some *Doctor Who* tales talk of a future where science will help us crash through these physical and intellectual barriers with progress in cybernetics and nanotech (or wetware, as it's sometimes called). Other tales imagine an enhanced human future through some genetic chance or accident. But the question common to both types of tales is this: Would human enhancement create supermen, or super-tyrants?

Doctor Who has kept its eye on science. The program writers have been aware of the way that science has driven the cutting-edge of tech so that ideas which once seemed fantastic now sit at the forefront of probability. But *Doctor Who* injects drama as this evolution of science reignites the long-standing nightmares of moral peril and as ethics struggle to keep pace with bleeding-edge tech. The chief dream of trans-humanism is the belief that humans will someday transform into different beings. Their abilities will evolve and become so different from the natural human condition as to merit the label of post-human beings. *Doctor Who* has provided us with plenty of examples of post-human beings more on the side of tyranny than democracy.

Doctor Who has produced writers who have conjured fantastic frameworks for tales of super-tyrants. Scientifically speaking, there is a natural difference between Darwin-induced narratives of "fitter" humans, and Lamarckian-inspired super-tyrants whose creative evolution is often instantaneous, and whose newfound powers are somehow passed on to their offspring. So, leaving alien races and armies to one side, here is a brief taxonomy of the super-tyrants of *Doctor Who*.

The Master

The most infamous renegade Time Lord and the Doctor's archenemy. The Master's ambitions were described in the 1976 story "The Deadly Assassin" as becoming "the master of all matter." His infamy has inspired descriptions such as "pure evil," according to the Eighth Doctor, "stone-cold brilliant," according to the Tenth Doctor, and "one of the most evil and corrupt beings [the] Time Lord race [had] ever produced," according to President Borusa.

Scientific aspect of super-tyranny: In the Tenth Doctor tale "The Sound of Drums," a flashback shows a young Master at the tender age of eight. During a Time Lord initiation ceremony, The Master is shown a gap in the fabric of the space-time continuum known as the Untempered Schism. From here the entire continuum can be seen. The Doctor suggests that gazing into the Schism caused some kind of shift in the Master's psyche, resulting in The Master becoming a kind of anti-Doctor, as anti-matter is to matter.

Davros

The ultimate mad scientist gone berserk, Davros was, of course, the creator of the Daleks. That single act already ranks him high on *Doctor Who's* list of super-tyrants. Not only that, but at one time Davros also led the Dalek race, heading up the Doctor's main enemy. A wheel-chaired megalomaniac, Davros is a boffin with a design on Whoniverse domination.

Scientific aspect of super-tyranny: It's long been the case that writers and filmmakers will use certain motifs to signify something slightly strange about a character. An audience learns more about a character from appearance than from pages of dialogue. This kind of recipe has been used many times in the portrayal of "mad" doctors such as Davros: unnamed scars, withered limbs, crazy eyes, curiously outsized foreheads, and, naturally, the wheelchair. Sigmund Freud wrote about this in his 1919 essay *The "Uncanny."* Humans possess specific psychological fears associated with missing parts of the human body. Freud wrote, "Dismembered limbs, a severed head, a hand cut off at the wrist . . . feet which dance by themselves—all these have something peculiarly uncanny about them, especially when, as in the last instance, they prove capable of independent activity in addition." While it's true that Freud's examples are drawn from a more ancient fantasy associated with witchy folk tales, his analysis still applies to the portrayal of Davros in *Doctor Who*.

House

House was the mysterious being at the heart of the Eleventh Doctor story "The Doctor's Wife." The Doctor described House as "a sort of sea urchin," which fed on TARDISes for their energy. House had an asteroid-sized body and a super-tyrant personality to match. House took pride in the fact that it had killed hundreds of Time Lords and enjoyed watching its victims suffer.

The Great Intelligence

Like House, the Great Intelligence was a formless entity, wandering in space. Having made many previous appearances, it encountered the Eleventh Doctor in 1892, when it took the form of snow. It found an ally in Doctor Simeon, and together they tried to create a race of ice-people. The Great Intelligence's exact nature was something of a mystery.

Scientific aspect of super-tyranny: Both House and the Great Intelligence are formless entities. This idea was first suggested by British astronomer Fred Hoyle, the man who was also famous for coining the phrase "big bang," though he didn't believe the theory himself. In an attempt to open up the idea of what extraterrestrial intelligence could be like, Hoyle's first novel was his 1957 book *The Black Cloud*. Here Hoyle explored ideas about disembodied intelligence, which takes the form of a cloud of interstellar matter, some five hundred million years old, and with which astronomers are ultimately able to communicate. Perhaps surprisingly, in the preface to *The Black Cloud*, Hoyle makes the point that "there is very little here that could not conceivably happen."

Has the Silence Visited the Human Past?

In the Eleventh Doctor story "Day of the Moon" (2011), the Doctor and his trusty companions attempt to lead humanity into a revolution against the Silence, a religious order of aliens who cannot be remembered after they are encountered. The Silence exist across the entire planet, and have the ability to place post-hypnotic suggestions in humans they encounter. The Doctor discovers that the Silence have been guiding humanity's evolution for millennia.

> "Some years ago, I came upon a legend, which more nearly fulfills some of our criteria for a genuine contact myth. It is of special interest because it relates to the origin of Sumerian civilization. Sumer was an early—perhaps the first—civilization in the contemporary sense on the planet Earth. It was founded in the fourth millennium BC or earlier. We do not know where the Sumerians came from. Their language was strange; it had no cognates with any known Indo-European, Semitic, or other language, and is understood only because a later people, the Akkadians, compiled extensive Sumerian-Akkadian dictionaries."
> —Carl Sagan and Iosif Shklovsky, *Intelligent Life in the Universe* (1966)

Silence Will Fall

The Silence are one of the greatest creations of *Doctor Who*. They are scarier than many of the program's past villains. They are creepy, and they are elusive. For their unsettling appearance, story writer Steven Moffat drew inspiration from Edvard Munch's famous 1893 expressionist painting *The Scream*, as well as from *Men in Black*. We learn that the Doctor first encountered the Silence in 1969 America, though he had been aware of their existence for some time. The Silence had been on Earth since the very start of civilization. Their mission was to secretly manipulate the evolution of humanity for their own enigmatic ends. In time we also find that the Silence are a religious movement composed of members of an alien-like humanoid species. The name of their religion is based on prophecy: when the oldest question in the Whoniverse is asked, silence will (or must) fall. As it's the Doctor who is allegedly prophesized to answer this question, which turns out to be the show's name itself ("Doctor who?"), the Silence aim to make his death a fixed point in space-time and prevent it. The Doctor is not so dumb, naturally. Though his death at Lake Silencio is indeed thought to be a "fixed point" of history, the Doctor faked his death, with River imprisoned for it so that the Silence will believe him dead.

Elusive Aliens

Steven Moffat's Silence are fascinating. They continue Moffat's preference for using simple psychological concepts to make his aliens and monsters more frightening. But they're bigger than that. The Silence are part of an interesting subgenre of sci-fi, where monsters and aliens are played far more subtly than before. By the late twentieth century, sci-fi had been imagining Darwinian aliens for over a century. But, in all that time, science still had precious little to say about the actual *form* of extraterrestrial life; that is, what aliens might actually *look* like. So some writers, ahead of the curve, had turned to more elusive explorations of alien contact. And *2001: A Space Odyssey* had raised science-fiction cinema to a new level with its celebrated and mature portrayal of mysterious, existential, and elusive aliens.

An intelligent writer like Moffat recognizes the scientific difference between alien and human—unlike in psychology as well as physiology and form. It's the same with the aliens in *2001*, created by writer Arthur C. Clarke and movie director Stanley Kubrick. In his book *Cosmic Connection: An Extraterrestrial Perspective*, American astronomer and SETI pioneer Carl Sagan confessed that Kubrick and Clarke had approached him about the best way to depict alien intelligence in their movie. (Sagan's science text *Intelligent Life in the Universe* had been something of a bible for Kubrick, as the film unfolded from concept into reality.)

Sagan's response was fascinating. He understood Kubrick's desire to portray aliens as humanoid. It was convenient, after all, and hardly uncommon. But Sagan pointed out that, since alien life forms were unlikely to bear any resemblance to Earthly life, to portray them as such would introduce "at least an element of falseness," into the film. Rather, Sagan suggested, the film should depict extraterrestrial super-intelligence, fitting for the Nietzschean theme of man's evolving into post-human superman. On attending the film's premiere, Sagan was "pleased to see that I had been of some help." And when pressed in an interview with *Playboy* in 1968, Kubrick hinted at the nature of the elusive aliens in *2001* by suggesting, given their long maturation, they had evolved from biological beings into "immortal machine entities," and then into "beings of pure energy and spirit," beings with "limitless capabilities and ungraspable intelligence." The elusive alien began to make "appearances" in other movies too. In Sagan's own book and movie *Contact*; in Arthur C. Clarke's novel *Rendezvous with Rama*; and in the Strugatsky brothers legendary novel *Roadside Picnic*.

The Diary of a Silent?

So Steven Moffat's Silence should be considered in this tradition of the elusive alien. Clarke and Kubrick had implied that their mysterious and existential aliens had given humanity a number of helpful hands up the evolutionary ladder, since the start of civilization. Moffat does something similar, though the Silence are far less benign than the elusive aliens in *2001*, as they secretly manipulate human evolution for their own ends. Here are some fanciful diary entries that a Silent may have made during his interventions in human history.

Date: Approximately 60,000 years ago
Spotted a bunch of human apes leaving Africa. Told them to head for the Fertile Crescent and tried to convince them (using wild and very large hand gestures) that the Crescent will be their cradle of civilization. Almost told them to Google it, but then remembered they haven't been helped to invent smartphones yet. They may have forgotten what I said anyhow.

Date: Approximately 30,000 years ago
The humans seem to like caves. I was hanging upside down like a bat in some caves in the south of France. Suddenly, I noticed some humans dabbling with primitive paint. One caveman wanted me to put my hand on the cave wall, so he could blow paint on it. I told him to use his own hand. Otherwise, those who follow will get quite the wrong idea about evolution. Think he later forgot I helped out, as I heard him grunting all the credit for himself. Ungrateful ape.

Date: January, 1892
Was knocking around Oslo, Norway, when I noticed a painter guy taking a walk next to the fjord. Sun was setting and the clouds were blood red. It was spooky and cool. Guy asked if I minded him painting me, as the scene was so dramatic. Was happy to oblige. It was so cold my ears fell off. Not sure he remembered me.

Date: July, 1969
As Neil Armstrong took the most important step for all mankind, he fell over. Luckily, I was there to pick him up. But no one remembered my good deed, and for some reason I didn't show up on the TV footage. Disappointed.

Date: 1995

Decided to try my hand at acting. Went along to the rehearsals for a new film, *Men in Black*, in which a couple of black-suited agents make people forget they've seen aliens. I consider myself perfect for the part. Film director looked terrified at first, but then admitted I was fit for the role. Didn't get the acting job, though. It's not that I was too thin. Or that the cameras seemed to ignore me standing there. I simply think the director forgot my rehearsal.

What's the *Doctor Who* Meme-Plex?

"It has sometimes been argued that the Cybermen (first appearance 1966) gave the world the 'cyber-' prefix to denote futuristic technology, though OED doesn't buy that, dating cybernetics from 1948 ('the theory or study of communication and control in living organisms or machines') and cyborg (cybernetic+organism) from 1960. Poor Cybermat, it tried so hard . . . "

—Tim Martin et al., "Doctor Who has changed the English language,"
The Daily Telegraph (2015)

"Meme: An element of a culture that may be considered to be passed on by non-genetic means, especially imitation."

—Susan Blackmore, *The Meme Machine (1999)*

Memes

Memetics is a relatively new science. English evolutionary biologist Richard Dawkins was the first to introduce the idea. He suggested that memes exist and evolve when he coined the word in his 1976 book *The Selfish Gene*. The notion gained momentum as a powerful metaphor in the debate on human evolution. The argument went something like this. First, biology. In biology, genes are replicators. Throughout evolution they have (mostly) faithfully transmitted the code of life from one generation to the next. They also allow natural selection to function by causing their carriers to survive, or propagate faster, than carriers of other genes. Second, culture. Dawkins was the first to argue that beliefs and ideas, memes, are also replicators. Memes are copied (mostly) faithfully from one mind to another. So the evolution of ideas is also governed by a kind of natural selection. Cultural change unfolds according to which memes replicate most speedily and effectively.

Dawkins's original intention was, in fact, to identify one fundamental principle that directs the development of evolving life anywhere in the Universe. As he wrote in *The Selfish Gene*, that fundamental principle was that "life evolves by the differential survival of replicating entities . . . I want to claim almost limitless power for slightly inaccurate self-replicating entities, once they arise anywhere in the Universe . . . given the right conditions, replicators automatically band together to create systems . . . that carry them around and work to favor their continued replication." And, in an attempt to offer another example of a replicator, in *The Selfish Gene* he posited the meme "to cut the gene down to size, rather than to sculpt a grand theory of human culture."

Dawkins's somewhat unintended idea is that cultural transmission is analogous to genetic transmission. Both, although essentially conservative, can give rise to a form of evolution, and "just as genes propagate themselves in the gene pool . . . memes propagate by imitation," in the meme pool. Dawkins's meme idea has been refined and developed in the years since *The Selfish Gene*. Perhaps the most interesting of these refinements is the idea of the meme-complex, or meme-plex. If an "idea-meme" is described as an entity capable of cultural transmission, then can we identify associations of memes, and does this association assist the survival of each of the participating memes? In *The Selfish Gene* Dawkins concludes:

> Co-adapted meme complexes evolve in the same kind of way as co-adapted gene complexes. Selection favors memes that exploit their cultural environment to their own advantage. This cultural environment consists of other memes, which are also being selected. The meme pool, therefore, comes to have the attributes of an evolutionary stable set, which new memes find it hard to invade.

Doctor Who Meme-Plex

Can we identify a meme-plex for *Doctor Who*? Can we spot a set of co-adapted memes that have not only been propagated by imitation, but have also made sure *Doctor Who* memes have been favored and selected in their cultural environment? Memes have certainly become a popular feature of social media in the twenty-first century. They come in the form of an image, a video, or a text-bite, typically humorous in nature, that are copied and spread rapidly over the Internet.

In some ways modern memes bear a close resemblance to the advertising slogans and movie taglines of old. A movie tagline, of course, is a concise piece of memorable text, which not only promotes the movie but also has a kind of dramatic effect on the recipient. Let's take a quick walk down memory lane and look at some of the best movie taglines. Sidney Lumet's terse jury-room drama *Twelve Angry Men* had the superb tagline, "Life is in their hands—death is on their minds." Martin Scorsese's study of existential loneliness in *Taxi Driver* had the tagline, "On every street in every city in this country, there is a nobody who dreams of being a somebody." And Peter Weir's satirical study of reality television *The Truman Show* boasted the jaunty rhyming tagline of "On the air and unaware."

Science fiction arguably has the most memorable taglines of all. Ridley Scott's famous *Alien* movie had the famous tagline of "In space no one can hear you scream." *Close Encounters of the Third Kind* had the tagline, "We are not alone." And *Star Wars: Episode IV - A New Hope* had cinema's most famous tagline, the genre-beating, "A long time ago in a galaxy far, far away . . . " But what spoken memes from *Doctor Who* could be co-adapted into a definitive meme-plex?

"EXTERMINATE!"

The battle cry of the Daleks has been terrifying audiences with their "EXTERMINATE" since the fifth episode in 1963. The Daleks and the meme turned *Doctor Who* into an overnight sensation. And yet both were very nearly exterminated before they made it to the screen. One BBC executive described Dalek creator Terry Nation's original script as "one of the worst things I've ever read," insisting "it can't go out." Luckily for the future *Doctor Who* meme-plex, the BBC didn't have any other stories ready to replace it. The rest is history. Essential to the meme-plex.

"Would you care for a jelly baby?"

A cunning ploy used by the Fourth Doctor to keep people waiting as he worked out what to do next. The First Doctor used a similar tactic with his own catchphrase, "Mm? What's that, my boy?" Not in the same league as the Dalek battle cry, so probably *not* in the meme-plex.

"YOU WILL BE DELETED!"

The clarion call of the Cybermen in the Tenth Doctor story *Rise of the Cybermen*, when the Cybermen, under businessman John Lumic, seek to "upgrade" all of humanity into Cybermen by placing their brains inside metal exoskeletons. A recent addition to the *Doctor Who* meme-plex.

"Allons-y!" and "Geronimo!"

"Allons-y" was a Tenth Doctor meme that captured the public imagination to such an extent that you could find it in online dictionaries associated with the French verb, aller, meaning to go. (allons-y means "let's go!") Unlike other memes in the plex, allons-y didn't survive too long. "Geronimo!" is something you shout when jumping from a great height, or as a general cry of excitement. The Eleventh Doctor used it in just that way, as if in some kind of reply to allons-y. (Although Steven Moffat blames Matt Smith for vocally inserting "Geronimo!" in different scripts where they hadn't been written in!) Like "allons-y," "Geronimo" doesn't make the plex cut.

"Bow ties are cool."
This Eleventh Doctor catchphrase was used by Steven Moffat as he was nervous of the Eleventh Doctor's bold new look and bow-tie fashion. Arguably, Moffat was right to be nervous. *Not* in the meme-plex.

"When I say run, RUUNNN!" and "Come along, Pond!"
There are interesting parallels between the Second and Eleventh Doctors. Neither was ever the warrior type. So when faced with danger the Second Doctor often said, "When I Say Run, RUUNNN!" and sensibly legged it, ensuring his companions were safe and sound. It's a tactic all the Doctors have used since, especially the Eleventh Doctor with his "Come along, Pond!" Used in no fewer than eight episodes, "Come along, Pond!" is well worth a mention as a possible candidate for the meme-plex. Also worth mentioning is the Eleventh Doctor's wonderful line, "I'm going to need a SWAT team ready to mobilize, street-level maps covering all of Florida, a pot of coffee, twelve Jammie Dodgers, and a fez." Neither make the meme-plex.

"Bigger on the inside!" and "Time and Relative Dimensions in Space"
When Peter Capaldi was interviewed by the BBC's *Radio Times* in 2017 about his departure from *Doctor Who*, he commented on the program's verbal memes. "I think there's loads of classic lines that are fun to say," Capaldi said. "I love saying 'Time and Relative Dimensions in Space' and 'Bigger on the inside' and 'They come from Skaro and will exterminate you.'" Capaldi continued, "I think they're part of the fabric of [this] country—they're in British popular culture which is nice but they will go on and on."

And so, taking the spoken memes "EXTERMINATE," "YOU WILL BE DELETED," "Bigger on the inside," and "Time and Relative Dimensions in Space" into a co-adapted set, we could argue we have our definitive *Doctor Who* meme-plex.

Do Sontarans Have Clone Girlfriends?

The Sontarans first made an appearance in the Third Doctor story "The Time Warrior" (1974). In this tale, the Sontaran Commander Linx crash-lands his spaceship in medieval England. He agrees to give futuristic weaponry to the warrior Irongron and his men, in exchange for Linx being given shelter to perform repairs on the damaged spaceship. The backstory of the Sontarans has not been revealed in the television series. It was a Doctor Who *role-playing game which claimed they were all descended from the genetic stock of General Sontar.*

> "That's all it is, there's nothing new about it, identical twins are clones. Anybody who objects to cloning on principle has to answer to all the identical twins in the world who might be insulted by the thought that there is something offensive about their very existence. Clones are simply identical twins. The only really deep reason people have for objecting to such a thing is that it just offends some deep-seated sense in people— what has been called the "yuk" reaction. It's irrational."
> —Richard Dawkins, BBC World Service interview (January 30, 1999)

Clones

Picture this moral dilemma. Your neighbor leaves their prize–winning goldfish at your home while they go on vacation. "Please look after Davros the goldfish while we're away. He's an odd fish, but we love him, all the same." Next morning, you awake to find that your cat has exhibited totally natural cat behavior and gobbled up Davros the goldfish during the night. Awkward. What to do? You can't replace Davros with just *any* old goldfish. He's such a one-off, with his designs on dominating not just *his* goldfish bowl, but also *all other* goldfish bowls beyond his own. Davros's owners are bound to notice if you try replacing him with a lesser fish. The simple solution to this moral dilemma? Clone Davros!

As we all know thanks to science fiction, a clone is an exact copy of an original; like an identical twin. If an organism is cloned, then the new version has exactly the same genes as the original. In fact, Chinese scientists created the world's first cloned fish in 1963, the same year *Doctor Who* was first on television. Science fiction then started explored ideas of human cloning. Movies about clones rather than copies or Doppelgängers became more common. The 1978 film *The Boys from Brazil* is a good example. Sadly, it's not about the World Cup, but it *is* about something equally fascinating: a mad doctor's attempt to clone Hitler and set up a fourth Reich (Hitler's regime, of course, being the Third Reich).

Doctor Who Clones

But *Doctor Who* covered clones even before *The Boys from Brazil*. The Sontarans were first created in a Third Doctor story, "The Time Warrior," way back in 1973. The Sontarans were a fictional race of alien humanoids, characterized by their ruthlessness and fearlessness of death. The origins of the Sontarans were not revealed in the television series, but a *Doctor Who* role-playing game claimed that they were descended from the genetic stock of General Sontar. They then used brand-new bioengineering tech to clone millions of duplicates of General Sontar and annihilate the non-clone population. Even Hitler didn't go *that* far. Sontar then named the new race after himself. He transformed the Sontarans into an expansionist, warrior society bent on universal conquest.

One wonders if the writers really think through some of the scientific and social implications of their plots. Take these cloned Sontarans, for example. Sure, they're a race of clones made for military purposes. And, being clones, they all look the same.

They're diminutive, corn-fed males, looking much like Mr. Potato Head in a jumpsuit. Like the actual Mr. Potato Head in *Toy Story*, you'd think this was bad news for getting girlfriends. Unless the Sontarans imitate the *Toy Story* plot and create diminutive, corn-fed females, looking much like *Mrs.* Potato Head in a jumpsuit.

But, romance aside, how do clones behave? Does having the same genes mean that a clone will have the same personality? Scientists have debated this for more than a century. It's known as the nature versus nurture debate. Some of our makeup is down to our genes (nature), and some down to the environment (nurture). One classic test is the study of identical twins that have been brought up in different families. Because they have the same genes, they will have the same physical features. But they will be nurtured in different ways because they have experienced a different upbringing. Although they are both important, most scientists think our personality is more down to our environment than our genes. Nurture wins out over nature. So if you're cloned like a Sontaran, your brother soldiers look like you, but they won't necessarily *behave* like you. Unless, of course, you're all raised in barracks in exactly the same environment.

Generally, clones don't last very long in the Whoniverse. Remember what the Sontarans did to Doctor companion Martha Jones? They linked Martha's clone to her original self via a mind transfer. But the clone couldn't survive after the link was severed. When the Fourth Doctor and Leela were cloned using a quick but unreliable method, the clones lived for only about ten minutes. And in the Twelfth Doctor story "Time Heist," Ms. Karabraxos believes that cloning herself is the only way to control her own security. And then incinerates her clones if they let her down.

Cloning Is Coming

Scientists began cloning animals first. Only a few years before an animal clone was made, US author Michael Crichton had written

Jurassic Park. Like Mary Shelley's *Frankenstein* before it, *Jurassic Park* was a warning not to mess around with nature. Written in 1990, with the famous movie coming three years later, *Jurassic Park* is about dinosaurs, of course. But the famous theme park monsters were cloned dinosaurs, and their creation leads to all kinds of unforeseen chaos.

Nevertheless, on July 5, 1996, the first-ever artificially cloned mammal was born. Dolly the sheep was made of DNA taken from a grown-up female sheep's teat. The breed they used was a Finn Dorset sheep, but the name Dolly was taken after the famous American country and western singer, Dolly Parton (who is also partly famous for her teats). Cloning animals may help preserve endangered species, and may also help research the cloning of human tissue.

Yet the main question *Doctor Who* is asking is this: If human cloning happens in the future, would it benefit everyone, or only those rich enough to afford it? There was no cost of cloning when, sporting an updated design, the Sontarans returned in the Tenth Doctor stories "The Sontaran Stratagem" and "The Poison Sky." The Sontarans planned to terraform the Earth into a new clone world, but their plans were averted. There was no mention of how they'd planned to pay for so many clone procedures.

Finally, on the question of a future of human cloning, it's worth considering the four laws of behavioral genetics, and how the laws impact identical twins. Law one says that all traits are partially heritable. Identical twins reared apart are more similar than fraternal twins reared apart. Law two states that the effect of the genes is larger than the effect of the shared environment. By the time they're adults, identical twins reared together are no more similar than identical twins reared apart. Law three says that a lot of variance in behavioral traits isn't down to either genes or the shared environment, which means that identical twins reared together aren't truly identical. And law four states that complex traits are usually shaped by many genes of small effect. For example, there's no common genes that affect IQ by, say, five points, but there are thousands of genes that affect IQ by a tiny fraction of a point. As *Doctor Who* has often pointed out, technology is a two-edged sword. It can be used for good, and bad. It's clear from the four laws of behavioral genetics that cloning will have consequences hard to predict, which means that there's plenty of scope for more Sontaran stories in the future.

Has the Doctor Ever Been Dolittle?

In the Twelfth Doctor story "Deep Breath" (2014), the Doctor and Clara investigate a series of incidents relating to apparent spontaneous combustion in Victorian London. The story starts with a dinosaur, which materializes in Victorian London and spits the TARDIS out onto the banks of the River Thames. A semi-delirious Doctor seems to understand what the dinosaur says.

Deep Breath

Does the Doctor have some of the skills of another famous Doctor in Dolittle? Doctor John Dolittle is the eponymous character in a series of books by Hugh Lofting. Doctor Dolittle lives in an English village where he cares for and talks to animals, rather than humans. Dolittle later becomes a naturalist, using his animal skills to better understand nature and the history of the natural world. In the Twelfth Doctor story "Deep Breath," the Doctor seems to have a similar talent. As Clara listens, the Doctor appears to translate the eerie howls of the Tyrannosaurus that is rampaging through Victorian London.

The episode makes one wonder about the efforts we mere earthlings have made to understand other animals on our planet. So far, it seems it's been all one-way traffic. We've made some effort to get animals to understand our languages, but not a lot of effort in trying to understand theirs. Consider gorillas and parrots. One particular gorilla, Koko, could grasp about two thousand words of spoken English. She hadn't evolved a vocal tract like humans, so she couldn't actually *speak* words. But this talented ape, who died on June 19, 2018, at the age of forty-six, gestured her thoughts using sign language. And that meant, if you include her native gorilla tongue, she spoke three languages, which is pretty impressive.

There's also the wonderful example of an African grey parrot called Alex. Alex could squawk around one hundred and fifty English words. This wordy birdie also showed that he could count up to six objects. He could make out different colors and shapes. He could combine words to make new meanings. And he could understand ideas like bigger, smaller, over, and under. On the night of his death in 2007, at the ripe old age of thirty-one, Alex's last words to his handler, Irene Pepperberg, were "You be good. See you tomorrow. I love you."

What's New, Pussycat?

Recently, scientists have begun to make more effort in understanding animal communication. But not without difficulty. Humans are meant to be the most intelligent species. You'd have thought we could learn to dolphin-speak better than dolphins could learn sign language. But dolphin-speak, it turns out, is not so easy to decode. Dolphins are smart animals. They form strong bonds, they socialize, and they seem to speak a code, a variety of clicks and whistles. But scientists have been trying to work out dolphin-speak for over half a century.

The main problem is working out what the "words" of dolphin-speak actually are. Are the dolphin whistles their version of human words? Or are clicks the dolphin's main way of speaking? Scientists have also noticed that dolphins use touch and posture in their language too. So they'll have to work out what that means before they can make any more progress.

Progress *has* been made in understanding the two-way conversations found throughout the animal kingdom. Like humans, animals engage in "turn taking" during two-way communications. "Turn taking" had previously long been said to be one of the main features that differentiated human language from the sounds made by our primate cousins. And yet new research suggests that many animal noises, from the rumbling of elephants to the chirps of naked mole rats, follow similar turn-taking rules.

New studies highlight timing as a significant aspect of conversational turn taking in both humans and animals. Some species were impatient chatterers. For example, some songbirds waited less than fifty milliseconds to respond during a conversation. (Humans pause for about two hundred milliseconds before replying in a conversation. Now, scientists have found that we aren't the only species who think it rude to interrupt.) At the much slower end of the scale, sperm whales converse in clicks, which had a gap of about two seconds between turns when chatting. Both European starlings and black-capped chickadees follow "overlap avoidance" during turn-taking conversations. It also seems that, like humans, animals are sensitive to rule-breaking during conversations. In some bird species, if chatter overlap happens, individual birds become silent or fly away, which implies that overlapping is treated as a kind of violation of the avian rules of turn-taking.

Human scientists are playing catch-up with the Twelfth Doctor in his conversation with the Tyrannosaurus in "Deep Breath." They have proposed a new framework for future animal research, which attempts to highlight key elements of human conversation. The aim is to facilitate large-scale, cross-species comparisons. In the future, such a framework will hopefully allow scientists to trace the evolutionary history of this incredible turn-taking behavior. It's also hoped that researchers will be able to address centuries-old questions about the origins of human language. Language, which is often thought of as the most distinctive human trait, is still mostly mysterious in terms of evolutionary theory. And, as turn-taking is thought to play a crucial role in the early development of human conversation, understanding it in other animals may help unpick the origin of human language.

Finally, how does the Twelfth Doctor understand the Tyrannosaurus in "Deep Breath"? There may be an element of telepathy—it *is* one of the Doctor's skills after all. And there are, of course, countless different spoken, written, and gestural languages used by millions of cultures throughout the space-time of the Whoniverse. And the Doctor has great expertise in linguistics, once claiming to be able to speak five billion languages. It could be that he was using the translation circuit of the TARDIS, though this is more often for the benefit of the Doctor's companions. Or it could be that the Doctor really *is* a bit Dolittle.

Are Humans Becoming Cybermen?

Cyberman: "Our brains are just like yours, except that certain weaknesses have been removed . . . "

Doctor: "The power cable generated an electrical field and confused their tiny metal minds. You might almost say they've had a complete 'metal' breakdown!"

Controller: "Activate them! The brain of this humanoid will be their target. Now!"

Doctor: "Well, they're a form of metallic life. They home on human brainwaves and attack."

—Various Cybermen quotes from *Doctor Who*

"We have no idea, now, of who, or what the inhabitants of our future might be. In that sense, we have no future. Not in the sense that our grandparents had a future, or thought they did. Fully imagined cultural futures were the luxury of another day, one in which "now" was of some greater duration. For us, of course, things can change so abruptly, so violently, so profoundly, that futures like our grandparents' have insufficient "now" to stand on. We have no future because our present is too volatile . . . we have only risk management. The spinning of the given moment's scenarios. Pattern recognition."

—William Gibson, *Pattern Recognition* (2003)

"We are programmed to be dissatisfied. Even when humans gain pleasure and achievements it is not enough. They want more and more. I think it is likely in the next two hundred years or so homo sapiens will upgrade themselves into some idea of a divine being, either through biological manipulation or genetic engineering or by the creation of cyborgs, part organic part non-organic. It will be the greatest evolution in biology since the appearance of life. Nothing really has changed in four billion years biologically speaking. But we will be as different from today's humans as chimps are now from us."

—Yuval Noah Harari, *The Daily Telegraph* (2015)

Twenty-First-Century Schizoid Man

This is the kind of cyborg chatter that you may see on social media in the twenty-first century: Are we evolving brains that are more like the brains inside the heads of the Cybermen? After all, scientists say our brains are busier than ever before. Every day we bombard our brains with facts, gibberish, and junk—all posing as information, so just struggling to keep up is a source of stress and exhaustion. Look at our smartphones. They've become somewhat like the Doctor's Sonic Screwdriver. Smartphones can now text, tweet, set calendar dates, calculate, spell-check, browse, email, and take high-resolution pictures. Not to mention Game Boy, voice recorder, weather forecaster, GPS, Facebook, and flashlight. Our smartphones have more computing power than the NASA mission that first put humans on the Moon in 1969. And our cyborg brains are becoming the same. We cram our brains 24–7. We text on the train, browse in the back of a baseball game, and dial-up a taxi when we're drunk. Many of us are obsessed with finding out what our friends are doing on Facebook, and it's all part of a twenty-first-century frenzy for "multitasking," and cramming data into every single spare moment of so-called "downtime."

What effect is all this having on our brains? Scientists say our brains are not wired to multitask. Our brains are best designed to focus just on one thing at a time. They don't like their time divided up into many things at once. So, when we're multitasking, it creates a kind of addictive feedback loop in the brain. And that rewards the brain for losing focus, and for constantly searching for more stimulation. This sounds alarmingly like a Cybermen plot. One can imagine a *Doctor Who* story where humans are slowly being taken over by machines. Gradually, we lose control. And eventually the Cybermen won't have to remove all thought and emotion from our brains. We would have done it for them!

Scholars Also Dream about Cybermen

Cyborgs are something of a science-fiction staple, of course. And, like the ruthless Darth Vader and the demonic Daleks in *Doctor Who*, cyborgs are among sci-fi's most famous villains. Indeed, *Doctor Who* got there pretty early. The program produced one of the most prominent cyborg incarnations in the form of the Cybermen.

Appearing first on British television way back in 1966, the Cybermen were a fictional race of cyborgs, a totally organic species to start with, who began to implant more and more artificial parts to help them survive. They are said to originate from Earth's twin planet Mondas (this is totally made up, as we don't actually have a twin planet, last time we looked). As the Cybermen like Victor Stone added more and more cyber parts, they became more coldly logical, calculating, and less human. As every emotion is deleted from their minds, they become less man, and more machine. Quite ingenious, really, as you can also use the idea of Cybermen as what humans may one day become, if we base all our decisions on cold calculation, and ignore our more human and emotional aspects.

Like *Doctor Who*, real-life scientists and writers are beginning to worry about our cyborg future. For example, in 2015 a professor at the Hebrew University of Jerusalem, Yuval Noah Harari, said the near-future cyborg fusion of man and machine would soon become the "biggest evolution in biology" since the emergence of life on Earth, four billion years ago. That's some claim!

Professor Harari hit the headlines in 2014 with his book, *Sapiens: A Brief History of Humankind*. Originally published in Hebrew under the title *A Brief History of Mankind*, Harari's work became a global phenomenon, attracting a legion of fans from Bill Gates and Barack Obama to Chris Evans and Jarvis Cocker, and available in nearly forty languages worldwide. Harari used the expertise gained from charting the history of humanity and turned his eyes to the near future. His verdict? Humans would evolve to become like gods, with power over death, and become as altered from the humans of today as we are from apes, maybe more so. Central to Harari's hypothesis is the idea that the human race is a striving species. We are driven by dissatisfaction, and we simply won't be able to resist the urge to "upgrade" ourselves, whether it's by genetic tinkering or augmenting tech. Yep, sounds like the Cybermen to me.

But Harari's cyborg future is not for everyone. Given the potential expense of the tech, he feels it's likely the imminent "go cyborg" option will be limited to the rich. And while the wealthy have the money to "go all Batman" and potentially live forever, the poor will simply die out. What led to humanity's dominance, now and in the future? According to Harari it is the human ability to invent "fictions," like *Doctor Who*.

Harari is talking about the kind of fictions that also hold society intact—fictions such as money, religion, and the idea of basic human rights, all of which have no basis in nature. Harari says that what enables humans to cooperate flexibly and exist in large societies is our imagination. Take religion. You can't convince a chimp to give you a banana with the promise it will get twenty more bananas in chimp heaven. It doesn't compute. But humans understand the promise. And money is the most successful story ever. Harari sees bankers as master storytellers. Finance ministers tell us that money is worth something. It isn't, of course. Even chimps know money is worthless. Harari also sees the concept of God as extremely important. Without religious myth you can't create society. God and religion are the most important fictions humans ever invented, according to Harari. And as long as humans believed they relied more and more on these gods they were controllable. Harari suggests the most interesting place in the world from a religious point of view is not the Middle East, but Silicon Valley. That's where they are developing a kind of techno-religion, just like the Cybermen. In Silicon Valley, they believe even death is just a technological problem to be solved. And what we see in the last few centuries, according to Harari, is humans becoming so powerful that they no longer need God, they just need technology.

Harari's conclusion is stunning for Whovians. He believes that in the next couple of centuries, just like a Cybermen script, we humans will upgrade ourselves into a kind of divine being. By engineering the creation of cyborgs, we will become post-human in what Harari calls the greatest evolution in biology since the dawning of life itself. Someone should get this guy to write an episode of *Doctor Who*!

Do Whoniverse Earthlings Carry Cybermen Organ Donor Cards?

In the Twelfth Doctor story "Deep Breath" (2014), the Doctor and Clara investigate a series of incidents relating to apparent spontaneous combustion in Victorian London. The investigation leads them to a deadly restaurant occupied by cyborgs. Clara also faces uncertainties about the Doctor's changed appearance after regeneration.

Spare Parts

Doctor Who has made itself clear over the years. You simply can't trust the Cybermen. They're always on the lookout for spare parts. Take the Twelfth Doctor story "Deep Breath." Beneath a restaurant in Victorian London, the Doctor and Clara find an underground chamber in which a Half-Faced Droid is sitting in a chair. Around him are pods containing other droids, all dressed up and looking like real Londoners. Then, the Doctor discovers the Half-Faced Droid is recharging in his chair. All is not what it seems. It's not, for once, a man building himself into a cyborg—but a robot who's adding on human flesh and body parts. The Doctor and Clara soon find out that the restaurant is a way of capturing random humans to harvest their flesh and organs. It seems that the droids need a constant supply of spares to replace their rotting organic parts.

Science fiction has something of a dodgy reputation with spare body parts. A little over two hundred years ago, *Frankenstein* was first published. For many folk the first-ever sci-fi book, *Frankenstein* famously features a creature that is entirely made up of spare body parts. Mary Shelley's tale is that of a scientist called Victor Frankenstein who becomes obsessed with making artificial life.

Of course, Victor wants his creature to be beautiful. But when he builds the man out of body parts, fresh from graveyards, the result is quite grotesque. (Not long after *Frankenstein,* scientists began to discover and dig up real monsters. The budding industrial revolution had unearthed the fossil record, as the great machines of the age turned over the earth. The world saw the start of dinosaur-mania, and the bones and body parts of extinct monsters were brought back to "life.")

A mere ten years after *Frankenstein* came a real-life scandal about human body parts: the body-snatcher murders in Scotland. These grisly killings were committed in Edinburgh, between 1827 and 1828, by William Burke and William Hare. Burke and Hare made their victims drunk, and suffocated them. Then, they sold the still-warm corpses of their victims to Edinburgh Medical College. They killed seventeen altogether, all in the name of education. Cold-blooded serial killers more than half a century before the days of Jack the Ripper. Their main customer was Professor Robert Knox. He used the corpses in the study of anatomy, which was blossoming at that time. You won't be surprised to hear that the city of Edinburgh was very fearful of the body-snatchers for a while.

Over one hundred years later, public horror about body parts had turned into amazement, as the first heart transplant was made. In 1967, Professor Christiaan Barnard made the first successful human-to-human transplant in Cape Town, South Africa. In the very same year, a sci-fi story predicted a future of black market body parts. The story was *The Jigsaw Man,* by American author Larry Niven. In the tale, Niven invented the fictional crime of organ-legging, meaning the piracy and smuggling of organs. The story imagines a future where the transplant of any organ is possible. Now, that *should* mean that life could be extended forever, but when the death rate goes down, the number of donors decreases. They needed fictional body-snatchers to keep the system going.

Cyber Parts

Doctor Who is missing a trick, here. Organ-legging seems like a whole new way that the Cybermen can make cyborgs. We've already seen that, with the help of Missy, they've tried the Nethersphere. Next up could be Cybermen Donor Cards and Cybermen who look like humans. In other words, Android-men. (Let's not forget the droids on the spaceship SS *Marie Antoinette*. It was the sister ship of the fifty-first-century spaceship SS *Madame de Pompadour*. As the Doctor said, that was another case of out-of-control repair droids using human body parts—this time to repair a spacecraft.)

The Cybermen know that human science and medicine are working hard on the question of aging. And a very splendid job is being done, to be sure. There was a time when life was nasty, brutish, and short. Even during the height of the grandeur of the Roman Empire, the average age of death was a mere thirty-five years. Today, the United Nations have predicted far out into the future. By the year 2300, they claim, life expectancy in most developed countries will be over one hundred years. The Cybermen can find little room for complaint there.

And yet, the Cybermen may point out, what happens if a human meets a mishap along the long way to their dotage? Perhaps the family robot dog goes really rabid, his circuits get infected with a virus, and he flips into "kill" mode. The human narrowly escapes death, but in the brawl K-9 manages to chomp off one of the human's arms. What then? Do you simply have to get used to the new nickname of "stumpy," or do you find a replacement body part?

The Body Pirates

In truth, however, we don't need to turn to Cybermen, or the future possibility of Organ Donor Cards. Body piracy is already here in fact. In 2016, legally donated human organs met less than 10 percent of world need, according to a report by the World Health Organization. In 2014, in the United States alone, the National

Kidney Foundation reported that 4,761 Americans died waiting for a transplanted kidney, with a further 3,668 dropping off the list as they became too sick to receive one. It's hardly surprising that some people turn to the black market to save their own lives. Estimates suggest that the illegal trade of all human body organs amounts to between $840 million and $1.7 billion annually, a figure which accounts for roughly 10 percent of transplanted organs. In a throwback to the dark days of Burke and Hare, these surgeries are often done by inadequately trained surgeons under unhygienic conditions.

So picture a future Earth, with the Cybermen in control. Most of the outside world has been contaminated by some kind of war with the Daleks. A community of humans, rescued from the toxic environment, believe they are living in a perfect but isolated colony. The reality is quite different and sinister, however. The colonists are actually clones. They are walking and talking spares. And their sole purpose is to provide organ donation for a new breed of Cybermen, who aim to become Android-men by adding on human flesh and body parts.

Index